International Vocational Education Bilingual Textbook Series
国际化职业教育双语系列教材

Professional Skill Training of Maintenance Electrician
维修电工职业技能训练

Ge Huijie   Chen Baoling
葛慧杰   陈宝玲   主 编

Beijing
Metallurgical Industry Press
2020

内 容 提 要

本书共分4个项目，主要内容包括：电工基本技能训练；三相异步电动机的基本控制技能训练；三相异步电动机的复杂控制技能训练；典型机床电气控制线路的故障检修。

本书既可作为职业院校机电相关专业的国际化教学用书，也可作为相关企业的培训教材和有关专业人员的参考书。

图书在版编目(CIP)数据

维修电工职业技能训练 = Professional Skill Training of Maintenance Electrician/葛慧杰，陈宝玲主编. —北京：冶金工业出版社，2020.9
国际化职业教育双语系列教材
ISBN 978-7-5024-8532-0

Ⅰ.①维… Ⅱ.①葛… ②陈… Ⅲ.①电工—维修—高等职业教育—双语教学—教材—汉、英 Ⅳ.①TM07

中国版本图书馆 CIP 数据核字(2020)第 169605 号

出 版 人 苏长永
地　　址　北京市东城区嵩祝院北巷39号　邮编　100009　电话　(010)64027926
网　　址　www.cnmip.com.cn　电子信箱　yjcbs@cnmip.com.cn
责任编辑　俞跃春　王 颖　美术编辑　郑小利　版式设计　孙跃红　禹 蕊
责任校对　李 娜　责任印制　李玉山
ISBN 978-7-5024-8532-0
冶金工业出版社出版发行；各地新华书店经销；三河市双峰印刷装订有限公司印刷
2020年9月第1版，2020年9月第1次印刷
787mm×1092mm　1/16；16印张；383千字；233页
52.00元

冶金工业出版社　投稿电话　(010)64027932　投稿信箱　tougao@cnmip.com.cn
冶金工业出版社营销中心　电话　(010)64044283　传真　(010)64027893
冶金工业出版社天猫旗舰店　yjgycbs.tmall.com
(本书如有印装质量问题，本社营销中心负责退换)

Editorial Board of International Vocational Education Bilingual Textbook Series

Director   Kong Weijun (Party Secretary and Dean of Tianjin Polytechnic College)

Deputy Director   Zhang Zhigang (Chairman of Tiantang Group, Sino-Uganda Mbale Industrial Park)

Committee Members   Li Guiyun, Li Wenchao, Zhao Zhichao, Liu Jie, Zhang Xiufang, Tan Qibing, Liang Guoyong, Zhang Tao, Li Meihong, Lin Lei, Ge Huijie, Wang Zhixue, Wang Xiaoxia, Li Rui, Yu Wansong, Wang Lei, Gong Na, Li Xiujuan, Zhang Zhichao, Yue Gang, Xuan Jie, Liang Luan, Chen Hong, Jia Yanlu, Chen Baoling

国际化职业教育双语系列教材编委会

主　任　孔维军（天津工业职业学院党委书记、院长）

副主任　张志刚（中乌姆巴莱工业园天唐集团董事长）

委　员　李桂云　李文潮　赵志超　刘　洁　张秀芳

　　　　　谭起兵　梁国勇　张　涛　李梅红　林　磊

　　　　　葛慧杰　王治学　王晓霞　李　蕊　于万松

　　　　　王　磊　宫　娜　李秀娟　张志超　岳　刚

　　　　　玄　洁　梁　奕　陈　红　贾燕璐　陈宝玲

Foreword

With the proposal of the 'Belt and Road Initiative', the Ministry of Education of China issued *Promoting Education Action for Building the Belt and Road Initiative* in 2016, proposing cooperation in education, including 'cooperation in human resources training'. At the Forum on China-Africa Cooperation (FOCAC) in 2018, President Xi proposed to focus on the implementation of the 'Eight Actions', which put forward the plan to establish 10 Luban Workshops to provide skills training to African youth. Draw lessons from foreign advanced experience of vocational education mode, China's vocational education has continuously explored and formed the new mode of vocational education with Chinese characteristics. Tianjin, as a demonstration zone for reform and innovation of modern vocational education in China, has started the construction of 'Luban Workshop' along the 'Belt and Road Initiative', to export high-quality vocational education achievements.

The compilation of these series of textbooks is in response to the times and it's also the beginning of Tianjin Polytechnic College to explore the internationalization of higher vocational education. It's a new model of vocational education internationalization by Tianjin, response to the 'Belt and Road Initiative' and the 'Going Out' of Chinese enterprises. Tianjin Polytechnic College and Uganda Technical College-Elgon reached a cooperation intention to establish the Luban Workshop to carry out vocational education cooperation on mechatronics technology and ferrous metallurgy technology major in 2019. The establishment of Luban Workshop is conducive to strengthen the cooperation between China and Uganda in vocational education, promote the export of high-quality higher vocational education resources, and serve Chinese enterprises in Uganda and Ugandan local enterprises. Exploring and standardizing the overseas operation of Chinese colleges, the expansion of international influences of China's higher vocational education is also one of the purposes.

The construction of 'Luban Workshop' in Uganda is mainly based on the EPIP (Engineering, Practice, Innovation, Project) project, and is committed to cultivating high-quality talents with innovative spirit, creative ability and entrepreneurial spirit. To meet the learning needs of local teachers and students accurately, the compilation of these international vocational skills bilingual textbooks is based on the talent demand of Uganda and the specialty and characteristics of Tianjin Polytechnic.

These textbooks are supporting teaching material, referring to Chinese national professional standards and developing international professional teaching standards. The internationalization of the curriculums takes into account the technical skills and cognitive characteristics of local students, to promote students' communication and learning ability. At the same time, these textbooks focus on the enhancement of vocational ability, rely on professional standards, and integrate the teaching concept of equal emphasis on skills and quality. These textbooks also adopted project-based, modular, task-driven teaching model and followed the requirements of enterprise posts for employees.

In the process of writing the series of textbooks, Wang Xiaoxia, Li Rui, Wang Zhixue, Ge Huijie, Yu Wansong, Wang Lei, Li Xiujuan, Gong Na, Zhang Zhichao, Jia Yanlu, Chen Baoling and other chief teachers, professional teams, English teaching and research office have made great efforts, receiving strong support from leaders of Tianjin Polytechnic College. During the compilation, the series of textbooks referred to a large number of research findings of scholars in the field, and we would like to thank them for their contributions.

Finally, we sincerely hope that the series of textbooks can contribute to the internationalization of China's higher vocational education, especially to the development of higher vocational education in Africa.

Principal of Tianjin Polytechnic College　Kong Weijun
May, 2020

序

  随着"一带一路"倡议的提出，2016年中华人民共和国教育部发布了《推进共建"一带一路"教育行动》，提出了包括"开展人才培养培训合作"在内的教育合作。2018年习近平主席在中非合作论坛上提出，要重点实施"八大行动"，明确要求在非洲设立10个鲁班工坊，向非洲青年提供技能培训。中国职业教育在吸收和借鉴发达国家先进职教发展模式的基础上，不断探索和形成了中国特色职业教育办学模式。天津市作为中国现代职业教育改革创新示范区，开启了"鲁班工坊"建设工作，在"一带一路"沿线国家搭建"鲁班工坊"平台，致力于把优秀职业教育成果输出国门与世界分享。

  本系列教材的编写，契合时代大背景，是天津工业职业学院探索高职教育国际化的开端。"鲁班工坊"是由天津率先探索和构建的一种职业教育国际化发展新模式，是响应国家"一带一路"倡议和中国企业"走出去"，创建职业教育国际合作交流的新窗口。2019年天津工业职业学院与乌干达埃尔贡技术学院达成合作意向，共同建立"鲁班工坊"，就机电一体化技术专业、黑色冶金技术专业开展职业教育合作。此举旨在加强中乌职业教育交流与合作，推动中国优质高等职业教育资源"走出去"，服务在乌中资企业和乌干达当地企业，探索和规范我国职业院校"鲁班工坊"建设和境外办学，扩大中国高等职业教育的国际影响力。

  中乌"鲁班工坊"的建设主要以工程实践创新项目（EPIP：Engineering，Practice，Innovation，Project）为载体，致力于培养具有创新精神、创造能力和创业精神的"三创"复合型高素质技能人才。国际化职业教育双语系列教材的编写，立足于乌干达人才需求和天津工业职业学院专业特色，是为了更好满足当地师生学习需求。

  本系列教材采用中英双语相结合的方式，主要参照中国专业标准，开发国际化专业教学标准，课程内容国际化是在专业课程设置上，结合本地学生的技术能力水平与认知特点，合理设置双语教学环节，加强学生的学习与交流能力。同时，教材以提升职业能力为核心，以职业标准为依托，体现技能与质量并重的教学理念，主要采用项目化、模块化、任务驱动的教学模式，并结合企业岗位对员工的要求来撰写。

本系列教材在撰写过程中，王晓霞、李蕊、王治学、葛慧杰、于万松、王磊、李秀娟、宫娜、张志超、贾燕璐、陈宝玲等主编老师、专业团队、英语教研室付出了辛勤劳动，并得到了学院各级领导的大力支持，同时本系列教材借鉴和参考了业界有关学者的研究成果，在此一并致谢！

最后，衷心希望本系列教材能为我国高等职业教育国际化，尤其是高等职业教育走进非洲、支援非洲高等职业教育发展尽绵薄之力。

天津工业职业学院书记、院长　孔维军

2020年5月

Preface

Tianjin Polytechnic College and Uganda Technical College-Elgon reached a cooperation intention to establish the Luban Workshop to carry out vocational education cooperation on mechatronics technology and ferrous metallurgy technology major in 2019. In order to strengthen the cooperation between China and Uganda in vocational education, the two colleges plan to compile a series of international vocational skills bilingual textbooks.

This book is one of the international vocational skills bilingual textbooks. Maintenance electrician is engaged in electrician, motor maintenance worker, electronic assembly worker, power generation and distribution, relay protection, factory electricity, CNC maintenance, home appliance maintenance, etc. It covers a wide range of work, including: layout, assembly, installation, commissioning, fault detection and troubleshooting, as well as maintenance of lines, fixtures, control devices, production lines and other electrical equipment in the factory. Therefore, the maintenance electrician position requires high comprehensive skills of talents.

According to the characteristics of mechanical and electrical students and practitioners, this book adopts the compilation style of project tasks. The content selection is divided into four projects from shallow to deep, which are composed of nine tasks. Each task consists of task description, task target, task analysis, task-related knowledge, bill of material, task implementation, task evaluation, task development, task summary. The content arrangement and classroom organization not only conform to the general cognitive law, but also close to the actual teaching environment. The implementation of classroom tasks based on the working process can fully guarantee the teaching effect.

This book is prepared by the maintenance electrician teaching team of Tianjin Vocational College of technology, and Ge Huijie is the chief editor in charge of Task 3.1 Preparation of Project 3, the compilation of auxiliary materials of the textbook and the revision of the unified draft of the textbook. Chen Baoling was in charge of Project 4.

Sun Xin is the editor in charge of Project 1. Geng Qingtao editor is responsible for Task 2.1 and Task 2.2 of Project 2. Wu Jun editor is responsible for Task 2.3 of Project 2 and Task 3.2 of Project 3.

Due to the limited level of editors, it is inevitable that there are some shortcomings in the book. Readers are welcome to make criticism and correction.

The editor
June, 2020

前　言

2019 年，天津工业职业学院与乌干达埃尔贡技术学院达成合作意向，共同建立"鲁班工坊"，就机电一体化技术专业、黑色冶金技术专业开展职业教育合作。双方计划编撰国际化职业教育双语系列教材。

本书是国际化职业教育双语系列教材之一。维修电工从事电工、电机维修、电子装配、发配电、继电保护、工厂用电、数控维修、家用电器维修等工作，其涵盖面十分广阔，维修电工的工作范围包括布局、组装、安装、调试、故障检测及排除，以及维修线路、固定装置、控制装置以及工厂内的生产线及其他电气类设备维修和维护。故维修电工岗位对人才的综合技能要求较高。

本书根据机电类学生和从业人员的特点，采用项目、任务的编写体例，内容选取上由浅入深分为 4 个项目、9 个任务。每个任务都包含任务描述、任务目标、任务分析、任务相关知识、材料清单、任务实施、任务评价、任务拓展、任务小结。内容编排和课堂组织上既符合普遍的认知规律，又贴近实际的教学环境。基于工作过程的课堂任务的实施可以充分保证教学的效果。

本书由天津工业职业学院维修电工教学团队编写，葛慧杰和陈宝玲担任主编，葛慧杰负责项目 3 中任务 3.1 以及教材附属材料的编写及教材统稿工作，陈宝玲负责项目 4 的编写，孙欣负责项目 1 的编写，耿青涛负责项目 2 中任务 2.1 和任务 2.2 的编写，吴君负责项目 2 中任务 2.3 和项目 3 中任务 3.2 的编写。

由于编者水平所限，书中不妥之处，敬请广大读者批评指正。

编者
2020 年 6 月

Contents

Project 1　Electrician Basic Skills Training 1

  Task 1.1　Electrical Safety Technology and First Aid for Electric Shock 1
    1.1.1　Task Description 1
    1.1.2　Task Target 1
    1.1.3　Task Analysis 1
    1.1.4　Task-related Knowledge 1
    1.1.5　Bill of Materials 11
    1.1.6　Task Implementation 11
    1.1.7　Task Evaluation 12
    1.1.8　Task Development 14
    1.1.9　Task Summary 16
  Task 1.2　Use of Digital Multimeter 16
    1.2.1　Task Description 16
    1.2.2　Task Target 16
    1.2.3　Task Analysis 16
    1.2.4　Task-related Knowledge 17
    1.2.5　Bill of Materials 23
    1.2.6　Task Implementation 24
    1.2.7　Task Evaluation 26
    1.2.8　Task Development 27
    1.2.9　Task Summary 30

Project 2　Basic Control Skill Training of Three-phase Asynchronous Motors 31

  Task 2.1　Cognition and Basic Test of Three-Phase Squirrel-cage Asynchronous Motor 31

· IX ·

2.1.1	Task Description	31
2.1.2	Task Target	31
2.1.3	Task Analysis	31
2.1.4	Task-related Knowledge	32
2.1.5	Bill of Materials	40
2.1.6	Task Implementation	40
2.1.7	Task Evaluation	41
2.1.8	Troubleshooting	43
2.1.9	Task Development	45
2.1.10	Task Summary	46

Task 2.2 Installation and Debugging of Starting Circuit of Three-Phase Asynchronous Motor ········ 47

2.2.1	Task Description	47
2.2.2	Task Target	47
2.2.3	Task Analysis	48
2.2.4	Task-related Knowledge	48
2.2.5	Bill of Materials	59
2.2.6	Task Implementation	60
2.2.7	Task Evaluation	61
2.2.8	Troubleshooting	63
2.2.9	Task Development	65
2.2.10	Task Summary	66

Task 2.3 Installation and Commissioning of Two-place Control Circuit of Three-Phase Asynchronous Motor ········ 67

2.3.1	Task Description	67
2.3.2	Task Target	67
2.3.3	Task Analysis	67
2.3.4	Task-related Knowledge	67
2.3.5	Bill of Materials	68
2.3.6	Task Implementation	68
2.3.7	Task Evaluation	70
2.3.8	Troubleshooting	71

2.3.9　Task Development …………………………………………… 73
2.3.10　Task Summary ……………………………………………… 73

Project 3　Complex Control Skill Training of Three-phase Asynchronous Motors ………………………………………………………… 75

Task 3.1　Installation and Debugging of Three-phase Asynchronous Motor Forward and Reverse Control Circuit ……………………………… 75

3.1.1　Task Description …………………………………………… 75
3.1.2　Task Target ………………………………………………… 75
3.1.3　Task Analysis ……………………………………………… 75
3.1.4　Task-related Knowledge …………………………………… 76
3.1.5　Bill of Materials …………………………………………… 80
3.1.6　Task Implementation ……………………………………… 81
3.1.7　Task Evaluation …………………………………………… 82
3.1.8　Troubleshooting …………………………………………… 83
3.1.9　Task Development ………………………………………… 86
3.1.10　Task Summary …………………………………………… 86

Task 3.2　Installation and Debugging of Three-phase Asynchronous Motor Step-down Starting Circuit …………………………………… 87

3.2.1　Task Description …………………………………………… 87
3.2.2　Task Target ………………………………………………… 87
3.2.3　Task Analysis ……………………………………………… 87
3.2.4　Task-related Knowledge …………………………………… 88
3.2.5　Bill of Materials …………………………………………… 92
3.2.6　Task Implementation ……………………………………… 92
3.2.7　Task Evaluation …………………………………………… 93
3.2.8　Troubleshooting …………………………………………… 95
3.2.9　Task Development ………………………………………… 96
3.2.10　Task Summary …………………………………………… 97

Project 4　Troubleshooting of Electrical Control Circuits of Typical Machine Tools ……………………………………………………… 98

Task 4.1　Fault Maintenance of Electrical Control Circuit of CA6140 Lathe ……… 98

- 4.1.1 Task Description ······ 98
- 4.1.2 Task Target ······ 98
- 4.1.3 Task Analysis ······ 98
- 4.1.4 Task-related Knowledge ······ 99
- 4.1.5 Bill of Materials ······ 104
- 4.1.6 Task Implementation ······ 105
- 4.1.7 Task Evaluation ······ 106
- 4.1.8 Troubleshooting ······ 107
- 4.1.9 Task Development ······ 108
- 4.1.10 Task Summary ······ 110

Task 4.2 Z35 Radial Drilling Machine Electric Control Line Fault Maintenance ······ 110

- 4.2.1 Task Description ······ 110
- 4.2.2 Task Target ······ 110
- 4.2.3 Task Analysis ······ 110
- 4.2.4 Task-related Knowledge ······ 111
- 4.2.5 Bill of Materials ······ 118
- 4.2.6 Task Implementation ······ 119
- 4.2.7 Task Evaluation ······ 120
- 4.2.8 Troubleshooting ······ 121
- 4.2.9 Task Development ······ 122
- 4.2.10 Task Summary ······ 124

References ······ 125

目 录

项目1 电工基本技能训练 ... 126

任务 1.1 电工安全技术与触电急救 ... 126
- 1.1.1 任务描述 ... 126
- 1.1.2 任务目标 ... 126
- 1.1.3 任务分析 ... 126
- 1.1.4 任务相关知识 ... 126
- 1.1.5 材料清单 ... 133
- 1.1.6 任务实施 ... 134
- 1.1.7 任务评价 ... 135
- 1.1.8 任务拓展 ... 136
- 1.1.9 任务小结 ... 137

任务 1.2 数字万用表的使用 ... 137
- 1.2.1 任务描述 ... 137
- 1.2.2 任务目标 ... 138
- 1.2.3 任务分析 ... 138
- 1.2.4 任务相关知识 ... 138
- 1.2.5 材料清单 ... 144
- 1.2.6 任务实施 ... 144
- 1.2.7 任务评价 ... 146
- 1.2.8 任务拓展 ... 147
- 1.2.9 任务小结 ... 149

项目2 三相异步电动机的基本控制技能训练 ... 151

任务 2.1 三相鼠笼式异步电动机的认识与基本测试 ... 151

2.1.1	任务描述	151
2.1.2	任务目标	151
2.1.3	任务分析	151
2.1.4	任务相关知识	152
2.1.5	材料清单	158
2.1.6	任务实施	159
2.1.7	任务评价	160
2.1.8	故障的排除	161
2.1.9	任务拓展	163
2.1.10	任务小结	164

任务 2.2　三相异步电动机的起动电路安装调试 165

2.2.1	任务描述	165
2.2.2	任务目标	165
2.2.3	任务分析	165
2.2.4	任务相关知识	166
2.2.5	材料清单	174
2.2.6	任务实施	175
2.2.7	任务评价	177
2.2.8	故障的排除	178
2.2.9	任务拓展	179
2.2.10	任务小结	181

任务 2.3　三相异步电动机两地控制线路安装调试 181

2.3.1	任务描述	181
2.3.2	任务目标	181
2.3.3	任务分析	181
2.3.4	任务相关知识	182
2.3.5	材料清单	182
2.3.6	任务实施	183
2.3.7	任务评价	184
2.3.8	故障的排除	185
2.3.9	任务拓展	187
2.3.10	任务小结	188

项目 3 三相异步电动机的复杂控制技能训练 ... 189

任务 3.1 三相异步电动机正反转控制线路安装调试 ... 189

- 3.1.1 任务描述 ... 189
- 3.1.2 任务目标 ... 189
- 3.1.3 任务分析 ... 189
- 3.1.4 任务相关知识 ... 189
- 3.1.5 材料清单 ... 192
- 3.1.6 任务实施 ... 194
- 3.1.7 任务评价 ... 195
- 3.1.8 故障的排除 ... 196
- 3.1.9 任务拓展 ... 198
- 3.1.10 任务小结 ... 199

任务 3.2 三相异步电动机降压起动线路安装调试 ... 199

- 3.2.1 任务描述 ... 199
- 3.2.2 任务目标 ... 199
- 3.2.3 任务分析 ... 199
- 3.2.4 任务相关知识 ... 200
- 3.2.5 材料清单 ... 203
- 3.2.6 任务实施 ... 203
- 3.2.7 任务评价 ... 205
- 3.2.8 故障的排除 ... 206
- 3.2.9 任务拓展 ... 207
- 3.2.10 任务小结 ... 208

项目 4 典型机床电气控制线路的故障检修 ... 209

任务 4.1 CA6140 型车床电气控制线路故障检修 ... 209

- 4.1.1 任务描述 ... 209
- 4.1.2 任务目标 ... 209
- 4.1.3 任务分析 ... 209
- 4.1.4 任务相关知识 ... 209
- 4.1.5 材料清单 ... 214

4.1.6	任务实施	214
4.1.7	任务评价	216
4.1.8	故障的排除	217
4.1.9	任务拓展	218
4.1.10	任务小结	219

任务 4.2　Z35 型摇臂钻床电气控制线路故障检修 220

4.2.1	任务描述	220
4.2.2	任务目标	220
4.2.3	任务分析	220
4.2.4	任务相关知识	220
4.2.5	材料清单	226
4.2.6	任务实施	227
4.2.7	任务评价	228
4.2.8	故障的排除	229
4.2.9	任务拓展	230
4.2.10	任务小结	232

参考文献 233

Project 1　Electrician Basic Skills Training

Task 1.1　Electrical Safety Technology and First Aid for Electric Shock

1.1.1　Task Description

Electricity will bring disaster to human beings as well as benefits to human beings. In the electrical accidents, the proportion of electric shock accidents is quite large. Therefore, it is necessary to learn the necessary knowledge of safe electricity use and emergency measures for electric shock.

1.1.2　Task Target

(1) Know the basic knowledge of electric shock.
(2) Master electrician's safe operation and safe use of electricity.
(3) Understand the common sense of first aid for electric shock and first aid methods on site.

1.1.3　Task Analysis

The task is divided into three sub tasks:
(1) Basic knowledge of electric shock;
(2) Safety operation procedures and safety electricity use knowledge of electricians;
(3) Common sense of first aid for electric shock and on-site first aid methods.

1.1.4　Task-related Knowledge

1.1.4.1　Basic Knowledge of Electric Shock

When the human body contacts with electricity, the voltage that will not cause any damage to the tissues of various parts of the human body (such as skin, heart, respiratory organs and nervous system) is called safety voltage.

The regulation of safe voltage value varies from country to country. For example, 24V for Holland and Sweden, 40V for America, 24V for France, 50V for DC, and 50V for Poland.

(1) Step voltage electric shock. When a phase grounding occurs to live equipment, the grounding current flows into the earth, showing different potentials at different points on the ground surface from the grounding point. The height of the potential is related to the distance from the grounding point, and the farther the distance, the lower the potential.

When people step on two points of the ground surface with different potential at the same time between feet, it will cause step voltage electric shock. In case of such a dangerous situation, close your feet and jump 20 meters away from the ground to ensure personal safety.

(2) Electric shock between phases. The so-called phase to phase electric shock means that when the human body is insulated from the earth, it contacts two different phase lines or the human body contacts two charged parts of different phases of the electrical equipment at the same time, at this time, the current passes through the human body from one phase line to another phase line, forming a closed circuit. This situation is called phase to phase electric shock. At this time, the human body is directly under the action of online voltage, which is more dangerous than single-phase electric shock.

The electric shock diagram of human body is shown in Figure 1-1.

Figure 1-1 Electric shock diagram of human body

(3) Lethal current. The minimum life-threatening current in a short period of time is called fatal current. When the current does not exceed 100mA, the main cause of fatal shock is ventricular fibrillation or asphyxia caused by current. Therefore, the current that causes ventricular fibrillation can be considered as fatal current.

(4) Risk and related factors of electric shock.

1) When the human body is shocked, the fatal factor is the current through the human body, not the voltage. But when the resistance is constant, the higher the voltage, the greater the current through the conductor. Therefore, the higher the voltage that the human body touches the electrified body, the greater the danger. But whether it is high voltage or low voltage, electric shock is dangerous.

2) The duration of electric current passing through human body is another important factor affecting the degree of electric shock injury. The longer the human body passes through the current, the lower the resistance of the human body, the greater the current of current conservation, and the more serious the consequences. On the other hand, for every contraction and expansion of the

human heart, there is about 0.1s interval, which is the most sensitive to current. If the current passes through the heart at this moment, even if the current is very small, it will cause heart tremor; if the current does not pass at this moment, even if the current is large, it will not cause heart paralysis. Therefore, if the current lasts for more than 0.1s, it will inevitably coincide with the most sensitive gap of the heart and cause great danger.

3) The way of electric current passing through human body is directly related to the degree of electric injury. If the current passes through the head of the human body, it will make the human immediately coma. If the current passes through the spinal cord, it will paralyze half of the human limbs. If the current passes through the heart, respiratory system and central nerve, it will cause nerve disorders or cause the heart to stop beating, interrupt the blood circulation of the whole body, and cause death. Therefore, the current path from hand to foot is the most dangerous. Second, the hand to hand current path, and then the foot to foot current path.

4) The frequency of electric current has a great influence on the degree of electric shock injury. 50Hz power frequency alternating current is more reasonable for the design of electrical equipment, but this frequency of current is also the most serious injury to the human body.

5) People's health condition, dry and wet skin of human body also have some influence on the degree of electric shock injury. Patients with heart disease, nervous system disease or tuberculosis are more seriously injured by electric shock than healthy people. In addition, the resistance of dry skin is large, the current passing through is small, and the resistance of wet skin is small, so the current passing through is large and the harm is great.

(5) Earth protection. Grounding protection, also known as protective grounding, is to connect the metal shell of electrical equipment with the grounding body, so as to prevent electric shock when the shell is electrified due to the insulation damage of electrical equipment.

(6) Contact potential, contact voltage, step potential and step voltage. When the grounding short-circuit current flows through the grounding device, the distributed potential is formed on the ground surface. The potential difference between the two points at the horizontal distance of 0.8m from the ground surface and the vertical distance of 1.8m along the equipment shell, frame or wall is called contact potential. The voltage that the human body bears when contacting the two points is called the contact voltage; the maximum potential difference between the center of the grounding grid mesh and the grounding body of the grounding grid is called the maximum contact potential, and the voltage that the human body bears when contacting the two points is called the maximum contact voltage. The potential difference between two points with a horizontal distance of 0.8m on the ground is called step potential. The voltage that the two feet of the human body bear when contacting the two points is called step voltage; the potential difference between the ground outside the grounding grid and the grounding body on the edge of the grounding grid at a horizontal distance of 0.8m is called the maximum step potential, and the voltage that the two feet of the human body bear when contacting the two points is called the maximum step voltage.

When an electric shock occurs, the current flowing through the human body is determined by the ratio of the electric shock voltage to the human body resistance. Human body resistance is not a fixed value. The resistance of all parts of the human body, except the cuticle, is the largest in the skin. When the human body is dry and undamaged, the resistance of the human body can be as high as 40000~400000Ω. If the skin is removed, the human body resistance can be reduced to 600~800Ω. But the skin resistance of human body is not fixed. When the skin is sweaty or damaged, the resistance will drop to about 1000Ω.

Perceptual current: when holding the power supply with your hand, the direct current that the palm of your hand feels hot, or the alternating current that the nerve is stimulated to feel slight tingling, is called perceptual current. The subjects put their hands on a small copper wire. The average perceived current of DC current is 5.2mA for men and 3.5mA for women.

Get rid of current: the current that can get rid of itself after electric shock is called get rid of current. According to the measurement results, the power frequency free current of men is 9mA, and that of women is 6mA. When the power frequency current of 18~22mA (get rid of the upper limit of the current) passes through the chest of the human body, the muscle reaction caused by it will make the person who gets the electric shock stop the current within the power on time, and the breathing can be recovered, and the adverse consequences will not be caused by the short-term breathing stop.

1.1.4.2 Electrical Safety Operation Procedures and Common Sense of Safe Use of Electricity

(1) Safety operation knowledge of electrician.

1) The metal shell of electric instrument and equipment shall have good grounding wire.

2) Use and replace the fuse according to the regulations. Do not increase the specification at will to avoid burning the instrument and equipment.

3) Rubber mat shall be laid on the operation site (ground).

4) Do not operate against rules and regulations, and form a good habit of one handed operation.

5) In case of special circumstances, firstly switch off the power supply.

6) Carry out safety inspection on equipment frequently to check whether there are exposed live parts and leakage. Exposed live wire ends must be wrapped with insulating materials in time. During the inspection, special electrical equipment shall be used, and under no circumstances shall it be identified by hand.

7) Install protective grounding or protective neutral. When the insulation of the equipment is damaged and the voltage jumps to its metal shell, the voltage on the shell shall be limited to a safe range, or the electrical equipment with damaged insulation shall be cut off automatically.

8) Use all kinds of safety tools correctly, such as insulating rod, insulating clamp, insulating gloves, insulating overshoes, insulating carpet, etc. And hang all kinds of warning signs and install necessary signal devices.

9) Install automatic leakage switch. When equipment leakage, short circuit, overload or personal electric shock occurs, the power supply will be cut off automatically to protect the equipment and personal.

10) When the power is cut off for maintenance and before connecting to the power supply, measures shall be taken to make other relevant personnel know, so as to prevent others from closing the power switch when someone is in the process of maintenance, or other personnel are working without knowing when connecting to the power supply, resulting in electric shock.

11) The shell of the low-voltage three-phase 380V power grid with the neutral point directly grounded adopts the protective neutral connection, and it is forbidden to adopt the protective grounding.

(2) Electrical safety technical measures. When working on electrical equipment and lines, especially in high voltage places, technical measures must be taken to ensure safety, such as power cut, power test, discharge, installation of temporary grounding wire, hanging of warning signs, installation of barriers, etc.

1) In case of power failure, all possible incoming lines shall be cut off and obvious disconnection points shall be provided. Special attention shall be paid to prevent reverse power transmission from low-voltage side by maintenance equipment, and measures shall be taken to prevent false closing.

2) For the power cut lines, the electroscope corresponding to the voltage level shall be used for power test.

3) The purpose of discharge is to eliminate the residual charge on the repaired equipment. Discharge can be operated by insulating rod or switch. Attention shall be paid to the discharge between the line and the ground and between the lines.

4) Install temporary ground wire. In order to prevent accidental power transmission and induction during operation, install temporary ground wire and short line on the maintenance equipment and line.

5) Hang warning signs and install them to block the power switch of the repaired equipment and lines. Lock them and hang the warning signs of 'no power transmission when someone works'. For the operation with partial power failure, the equipment without power failure whose safety distance is less than 0.7m shall be installed with temporary barriers and hung with 'stop, high voltage danger' signs.

The electrical safety technical measures is shown in Figure 1-2.

Figure 1-2 Electrical safety technical measures

1.1.4.3 Common Sense of First Aid for Electric Shock And on-site First Aid Methods

After the electric shock is separated from the power supply, rescue shall be carried out on the spot immediately. The meaning of 'immediate' is to seize every minute and second without delay. The meaning of 'on the spot' is not to wait for the doctor's arrival passively, but to send someone to inform the medical staff to go to the scene and make preparations to send the person who gets electric shock to the hospital at the same time of carrying out correct rescue on the scene.

The first aid measures for electric shock is shown in Figure 1-3.

Figure 1-3 First aid measures for electric shock

(1) On site rescue measures.

1) Rescue measures for those who are not unconscious due to electric shock. If the injured person is not seriously injured and is still conscious, but has palpitations, dizziness, cold sweat, nausea, vomiting, numbness of limbs, general weakness, or even coma for a time, but does not lose consciousness, the person should be allowed to lie down in a warm and ventilated place for rest, and sent for close observation. Meanwhile, the doctor should be invited to come or send to the hospital for diagnosis and treatment.

2) Rescue measures for patients who have lost consciousness. If the person has lost consciousness, but his breathing and heart rate are normal, he/she should lie on the ground comfortably, untie his/her clothes to facilitate breathing, do not surround people around, keep the air flowing, keep warm in cold days, and immediately ask the doctor to come or send him/her to the hospital for consultation. In case of dyspnea or cardiac arrhythmia of the person with electric shock, artificial respiration or extrathoracic cardiac compression should be performed immediately.

3) First aid measures for 'feigned dead'. If the person with electric shock presents the phenomenon of 'pseudo death' (the so-called electric shock), there may be three kinds of clinical symptoms: first, the heart stops, but still can breathe; second, the breath stops, but the heart still exists (the pulse is very weak); third, the breath and the heart have stopped. The methods of judging the symptoms of 'feign death' are 'look', 'listen' and 'try'. 'Look' is to observe

whether the chest and abdomen of the person who is shocked have ups and downs; 'listen' is to use the ear to close to the mouth and nose of the person who is shocked, and listen to whether he has breath sound; 'test' is to use the hand or small paper strip to test whether there is breath air flow in the mouth and nose, and then use two fingers to gently press one side (left or right) of the carotid artery beside the laryngeal node depression to check whether there is pulsation. If the results of 'watching', 'listening' and 'testing' show that there is neither breath nor carotid pulsation, it can be determined that the person with electric shock stops breathing or the heartbeat stops or both.

(2) Cardiopulmonary resuscitation. When it is judged that the person with electric shock stops breathing and heartbeat, it shall be rescued on the spot according to cardiopulmonary resuscitation immediately. The so-called cardiopulmonary resuscitation is to support the three basic measures of life, namely, unobstructed airway; mouth-to-mouth (nose) artificial respiration; external chest compression (artificial circulation).

1) Unobstructed airway.

① Remove foreign matters in the mouth. Make the person lying on his back in a flat and hard place, and quickly untie his collar, scarf, leotard and trouser belt. If there are foreign bodies such as food, denture and blood clot in the mouth of the person with electric shock, turn his body and head at the same time, quickly insert them from the corner of the mouth with one finger or two fingers, and take out the foreign bodies from them. During the operation, pay attention to prevent the foreign bodies from being pushed into the throat.

② Using the method of raising jaw with head up to clear the airway. During the operation, the rescuer puts one hand on the forehead of the person who is shocked, the fingers of the other hand lift up the chin and jaw of the person, the two hands push the head back together, the tongue root will lift naturally, and the airway will be unblocked. In order to make the head of the person with electric shock recline, appropriate thickness of articles can be padded under the neck, but it is strictly prohibited to use pillows or other articles to pad under the head of the person with electric shock, because the head raised and reclined forward will block the airway, reduce the blood flow to the brain during chest compressions, or even completely disappear.

The schematic diagram of head up and jaw up method is shown in Figure 1-4.

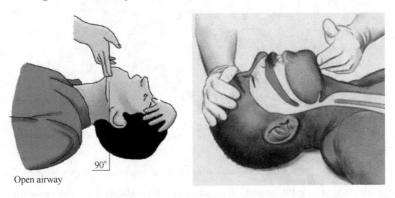

Figure 1-4　Schematic diagram of head up and jaw up method

2) Mouth to mouth (nose) artificial respiration. After the rescuer completes the airway unobstructed operation, he should immediately carry out mouth to mouth or mouth to nose artificial respiration to the person who is shocked. Mouth to nose artificial respiration is used to close the mouth of people with electric shock.

First, blow with a big mouth to stimulate pacing, the rescuer squats and kneels on the left or right side of the person who is shocked; hold the nose wing with the finger of the hand on the forehead of the person who is shocked, and gently hold the chin with the index finger and middle finger of the other hand; after the rescuer inhales deeply, close the mouth to mouth with the person who is shocked, and blow with a big mouth twice in a row without air leakage, 1~1.5s each time; then test the neck of the person who is shocked with the finger. If there is pulsation in the artery, if there is still no pulsation, it can be judged that the heartbeat has stopped. External chest compression should be carried out at the same time of artificial respiration.

After two times of carotid pulse test, the patients were immediately transferred to the normal stage of mouth-to-mouth artificial respiration. The normal blowing frequency is about 12 times per minute. The normal operation posture of mouth to mouth artificial respiration is as above. However, it is not necessary to blow too much air to avoid stomach expansion. If the person with electric shock is a child, it is better to blow less air to avoid rupture of alveoli. When rescuing a person for breath, the nose or mouth of the person who is shocked should be relaxed to let him breathe automatically by the elasticity of his chest. When blowing and relaxing, pay attention to whether the chest of the person with electric shock has undulating breathing action. If there is a large resistance during blowing, it may be that the head is not tilted back enough, so it should be corrected in time to keep the airway unblocked. If the teeth are closed tightly in case of electric shock, mouth to nose artificial respiration can be changed. When blowing, the lips of the person with electric shock shall be tightly closed to prevent air leakage.

The schematic diagram of mouth to mouth artificial respiration is shown in Figure 1-5.

Figure 1-5 Schematic diagram of mouth to mouth artificial respiration

3) Chest compression. Chest compressions are the first-aid methods to recover the heart beat of the person with electric shock by means of manpower. Its validity lies in choosing the right press position and taking the right press posture.

① To determine the correct pressing position:

First, the forefinger and middle finger of the hand are upward along the lower edge of the right costal arch of the electrocuted person, and the midpoint of the joint between the rib and sternum is found.

Second, the two fingers of the right hand are parallel, the middle finger is placed at the middle point of the notch (the base of the xiphoid process), the index finger is placed at the lower part of the sternum, and the palm root of the other hand is placed on the sternum next to the upper edge of the index finger, which is the correct pressing position.

The correct position of chest compression is shown in Figure 1-6.

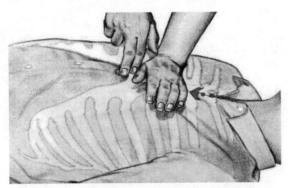

Figure 1-6 Correct position of chest compression

② Correct pressing posture:

First, make the person lying on his back care about the hard place and untie his clothes. The position of lying on his back is the same as that of mouth to mouth (nose) artificial respiration.

Second, the rescuer stands or kneels beside one shoulder of the person who is shocked. The two shoulders are directly above the sternum of the person who is shocked. The two arms are straight, the elbow joint is fixed and does not bend. The two palms are overlapped, the fingers are raised and do not touch the chest wall of the person who is shocked.

Third, with the hip joint as the fulcrum, the normal adult sternum was vertically depressed 3~5cm by using the gravity of the upper body.

Fourth, when it reaches the required level, it shall be relaxed immediately, but the palm root of the rescuer shall not leave the chest wall of the person who is shocked. The effective sign of pressing is that the carotid pulse can be touched during pressing.

The correct chest compression is shown in Figure 1-7.

③ Proper press frequency:

First, chest compression should be carried out at a uniform speed. The operation frequency should be 80 times per minute, each time including a cycle of pressing and relaxing. The time of pressing and relaxing is equal.

Second, when external chest compression and mouth-to-mouth (nose) artificial respiration are carried out simultaneously, the operation rhythm is: in case of single rescue, blow twice (15 : 2)

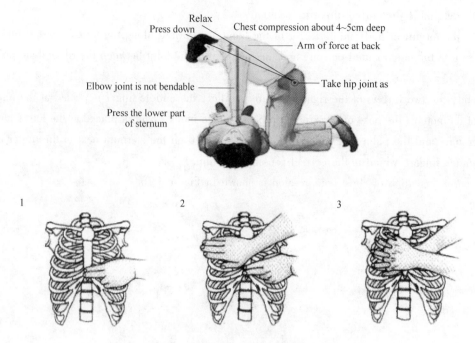

Figure 1-7 Correct chest compression

after every 15 times of compression, and repeat; in case of double rescue, blow once (15 : 1) after every 15 times of compression by another person, and repeat.

(3) Precautions in field rescue.

1) In the process of rescue, it is necessary to re judge the person who gets electric shock in time. After pressing and blowing for 1 minute (equivalent to four 15 : 2 cycles for single rescue), the method of 'look, listen and try' should be used to judge whether the person who has been electrocuted recovers natural breathing and heartbeat within 5~7s. If it is determined that the electrocuted person has carotid pulsation but still has no breath, the external chest compression can be suspended, and the mouth-to-mouth artificial respiration can be carried out twice, and then the breath can be blown every 5s (equivalent to 12 times per minute). If the pulse and breath are still not recovered, the cardiopulmonary resuscitation should be continued. In the process of rescue, it is necessary to judge the breathing and pulse condition of the person with electric shock every few minutes with the method of 'look, listen and try', and the time of each judgment shall not exceed 5~7s. The on-site personnel shall not give up the on-site rescue before the medical personnel come to take over the rescue.

2) Precautions for transferring the wounded with electric shock in the process of rescue:

① Cardiopulmonary resuscitation should be carried out on site, and the wounded should not be moved at will for convenience. If it is necessary to move, the interruption time of rescue should not exceed 30s.

② When moving or sending an electric shock to the hospital, a stretcher shall be used and a board shall be placed on the back of the person, and the person shall not be carried curled up.

The rescue shall be continued during the transfer, and shall not be interrupted before the medical personnel take over the rescue.

③ Conditions shall be created to wrap the plastic bag with ice chips around the head of the wounded in a cap shape, so as to expose the eyes, reduce the brain temperature, and strive for the recovery of the heart, lung and brain of the electrocuted person.

④ If the heart rate and breath of the person with electric shock have recovered after rescue, cardiopulmonary resuscitation can be suspended. But in the early stage of recovery of heartbeat and respiration, it is still possible to stop suddenly again. The rescuer should be closely monitored, not be paralyzed, and be ready to rescue again at any time. At the beginning of recovery, the person with electric shock is often delirious, absent-minded or restless, so he should try to be quiet.

1.1.5 Bill of Materials

The table of the bill of materials is shown in Table 1-1.

Table 1-1 Bill of materials

No.	Name	Quantity
1	XK-SX2C advanced maintenance electrician training platform	1
2	One maintenance electrician training component (XKDT11)	1
3	One maintenance electrician training component (XKDT12A)	1
4	One warning sign to prevent electric shock	1
5	One simulation teaching model of chest compression	1

1.1.6 Task Implementation

The table of the mission statement is shown in Table 1-2.

Table 1-2 Mission statement

( ) mission statement					
1. According to the requirements, complete the required devices and complete the following table					
No.	Device name	Number of devices	Device parameters	Device measurement	Remarks

Continued Table 1-2

( ) mission statement
2. Job description
3. Operation process record
4. Task summary

1.1.7 Task Evaluation

The table of the scoring table is shown in Table 1-3.

Table 1-3 Scoring table

( ) scoring table					
Device inventory and measurement (10 points)					
No.	Key inspection contents	Scoring criteria	Weighting	Score	Remarks
1	Device inventory	Deduct 1 point for each error in device counting	5		
2	Device placement	Deduct 1 point for each item with wrong device placement	5		
Subtotal					

Continued Table 1-3

( ) scoring table

Correct use of safety electricity warning signs (10 points)

No.	Key inspection contents	Scoring criteria	Weighting	Score	Remarks
1	Code for safe operation of power off and transmission	Deduct 1 point for one mistake	5		
2	Usage of safety warning signs	Deduct 1 point for one mistake	5		
		Subtotal			

Correct operation of mouth to mouth artificial respiration (25 points)

No.	Key inspection contents	Scoring criteria	Weighting	Score	Remarks
1	Position of teaching aids	Whether the posture is normal	5		
2	Operation steps	Is it the procedure complete	20		
		Subtotal			

Correct operation of chest compression (25 points)

No.	Key inspection contents	Scoring criteria	Weighting	Score	Remarks
1	Position of teaching aids	Deduct 1 point for one mistake	5		
2	Press position	Deduct 1 point for one mistake	10		
3	Press posture	Deduct 1 point for one mistake	10		
		Subtotal			

First aid treatment for electric shock on site (20 points)

No.	Key inspection contents	Scoring criteria	Weighting	Score	Remarks
1	The judgment of the person who gets electric shock	Understand or not	5		
2	Transfer the wounded	Understand or not	5		
3	Treatment of the wounded after recovery	Understand or not	10		
		Subtotal			

Professional quality (10 points)

No.	Key inspection contents	Scoring criteria	Deduction	Score	Remarks
1	Power safety awareness	Deduct 5 points at a time until all points are deducted			
2	Standard operation	Deduct 1 point at a time until all points are deducted			
3	Team work	As appropriate			
4	Clean working position	As appropriate			
		Total			

1.1.8 Task Development

Electrical fire refers to the disaster caused by combustion caused by electrical reasons. Short circuit, overload, leakage and other electrical accidents may cause fire. The direct causes of electrical fire are equipment defects, improper construction and installation, poor electrical contact, high temperature caused by lightning and static electricity, arc and spark. It is the environmental condition of electrical fire to store flammable and explosive materials around.

Direct causes of electrical fire:

(1) There is a short circuit fault in the equipment or line. Electrical equipment will cause short-circuit fault due to insulation damage, circuit disrepair, carelessness, operation error and unqualified equipment installation. The short-circuit current of electrical equipment can reach tens or hundreds of times of normal current. The heat generated (proportional to the square of current) is caused by the temperature rise exceeding the ignition point of itself and surrounding combustibles, resulting in fire.

(2) Overload causes overheating of electrical equipment. It is unreasonable to select the line or equipment, the load current of the line exceeds the rated safe carrying capacity of the conductor, the electrical equipment is overloaded for a long time (exceeding the rated load capacity), which causes the line or equipment overheating and leads to fire.

(3) Poor contact causes overheating. If the connector is not connected firmly or tightly, the pressure of the moving contact is too small, the contact resistance is too large, and the contact part overheats, causing fire.

(4) Poor ventilation and heat dissipation. The high-power equipment is lack of ventilation and heat dissipation facilities or the ventilation and heat dissipation facilities are damaged, causing overheating and causing fire.

(5) Improper use of electrical appliances. For example, the electric stove, iron and soldering iron are not used as required, or the power supply is forgotten to be disconnected after use, resulting in overheating and fire.

(6) Electric sparks and arcs. Some electrical equipment can produce electric spark and arc in normal operation, such as the opening and closing operation of large capacity switch and contactor contact, which will produce electric arc and spark. The temperature of electric spark can reach thousands of degrees, which can be ignited in case of combustible materials and explode in case of combustible gas.

In every place of daily life and production, there are a lot of inflammable and explosive substances, such as liquefied petroleum gas, gas, natural gas, gasoline, diesel oil, alcohol, cotton, hemp, chemical fiber fabric, wood, plastic, etc. in addition, some equipment itself may produce inflammable and explosive substances, such as the decomposition and gasification of insulating oil of equipment under the action of electric arc, spraying a large amount of oil mist and combustible gas acid battery discharges hydrogen and forms explosive mixture. Once these inflammable and explosive environments encounter the fire source caused by electrical equipment and circuit failure, they will immediately ignite and burn.

The protection measures for electrical fire are mainly devoted to eliminating hidden dangers and improving the safety of electricity use. The specific measures are as follows.

(1) Correctly select protective devices to prevent electrical fire. Heat insulation, heat dissipation, forced cooling and other structures shall be adopted for the equipment that may produce electric heating effect under normal operation conditions, and the use of heat-resistant and fireproof materials shall be emphasized. Automatic power-off protection including short circuit, overload and leakage protection equipment shall be set according to the specified requirements. The electrical equipment and lines shall be correctly set with grounding and neutral protection, and lightning arresters and grounding devices shall be installed for lightning protection. The electrical equipment shall be correctly designed and selected according to the use environment and conditions. In the harsh natural environment and places with conductive dust, products with anti insulation aging function shall be selected, or corresponding measures shall be added; explosion-proof electrical products must be used in inflammable and explosive places.

(2) Correctly install electrical equipment to prevent electrical fire. Select the installation location reasonably. For the explosion hazardous area, the electrical equipment should be installed outside the explosion hazardous area or the place with less explosion risk. Switches, sockets, fuses, electric heating appliances, welding equipment and motors shall be kept away from inflammables or inflammable building components as far as necessary. Inflammables shall not be stacked under the crane sliding contact line. Outdoor power transformation and distribution equipment shall not be set in places where combustible dust or fiber are easy to deposit.

Keep necessary fire distance. For electrical equipment that can generate electric arc or electric spark in normal operation, arc extinguishing materials shall be used to isolate them completely, or to keep sufficient distance between them and materials that may be ignited, arc resistant materials or materials that may cause fire, so as to ensure safe arc extinguishing.

When installing and using electrical equipment with local heat focus or heat concentration, a sufficient distance must be kept between the direction of local heat focus or heat concentration and flammable materials to prevent ignition.

The protective barrier materials around the electrical equipment must be able to withstand the high temperature generated by the electrical equipment (including in case of failure). Nonflammable and flame-retardant materials shall be selected according to specific conditions or fire-retardant coating shall be sprayed on the surface of combustible materials.

Keep the normal operation of electrical equipment to prevent electrical fire. Correct use of electrical equipment is the premise to ensure the normal operation of electrical equipment. Therefore, the electrical equipment shall be operated according to the provisions of the equipment operation manual. Strictly implement the operating procedures. Keep the voltage, current and temperature rise of electrical equipment within the allowable value. Keep all conductive parts connected reliably and grounded well. Keep the insulation of electrical equipment in good condition, keep the electrical equipment clean and keep good ventilation.

1.1.9　Task Summary

The safe use of electricity includes three aspects: the safety of power supply system, the safety of electrical equipment and personal safety, which are closely linked. The failure of the power supply system may lead to the damage of the electrical equipment or personal injury accidents, and the electrical accidents may also lead to local or large-scale power failure, even cause serious social disasters.

In the process of using electricity, special attention must be paid to electrical safety. If there is slight paralysis or negligence, it may cause serious personal electric shock accident, or cause fire or explosion, which will bring great loss to the country and people.

Task 1.2　Use of Digital Multimeter

1.2.1　Task Description

In various working environments of electrical engineering, it is often necessary for staff to monitor and measure the equipment, which requires us to use various measuring instruments. The digital multimeter is a relatively simple measuring instrument. Its characteristics include: high input impedance, small input capacitance, so it has little influence on the circuit to be tested; wide measurement frequency range. Generally, it can range from dozens of Hz to several megahertz; High sensitivity; large voltage measurement range, as low as microvolt level or millivolt level, as high as thousands of volts and thousands of volts; direct indication, digital multimeter can read the electric quantity value directly from the LCD screen when measuring, which is convenient and intuitive; full function, with all functions of ordinary multimeter and other additional functions (such as measuring transistor and other relevant parameters).

This task focuses on the correct use of digital multimeter. Starting from the measurement methods of voltage, resistance, current, diode, triode and MOS FET of the digital multimeter, let us better master the measurement methods of the multimeter.

1.2.2　Task Target

(1) Master the use of digital multimeter.

(2) Understand the precautions of Digital Multimeter in use.

1.2.3　Task Analysis

Digital multimeter is a kind of multi-purpose electronic measuring instrument, generally including ammeter, voltmeter, ohmmeter and other functions, sometimes also known as multimeter, multi meter, multi meter, or three meter. The digital multimeter has portable devices for basic fault diagnosis, as well as devices placed on the workbench, some of which can reach a resolution of seven or eight bits. DMM is the electronic instrument used in electrical measurement. It can have many special functions, but its main function is to measure the voltage, resistance and current. As a modern multi-purpose electronic measuring instrument, it is mainly used in physical, electrical, electronic and other measuring fields.

The task is divided into eight sub tasks:

(1) Voltage measurement;

(2) Current measurement;

(3) Resistance measurement;

(4) Diode measurement;

(5) Buzzer on-off test;

(6) Triode measurement;

(7) MOS FET measurement;

(8) Precautions for using digital multimeter.

1.2.4 Task-related Knowledge

The working principle of the digital multimeter is to amplify the measured signal, and then process it digitally, and finally display the measurement results in digital form by the digital meter. It has high measurement accuracy, high resolution, high voltage sensitivity, multiple measurement types, complete functions, strong overload ability, good anti-interference, small volume, light weight and reliability High performance, and because of the use of digital form to display the measurement results, making the reading fast and convenient, and can fundamentally eliminate the reading error caused by parallax, so it is widely used in electrical, electronic, communication, scientific research and home appliance industry.

The appearance and function diagram of DMM is shown in Figure 1-8.

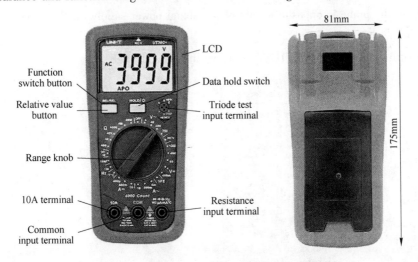

Figure 1-8 Appearance and function diagram of DMM

The symbol description of digital multimeter is shown in Figure 1-9.

		Symbol	Function
⎓	DC	V~	AC voltage measurement
~	AC	V⎓	DC voltage measurement
≂	DC/AC	A~	AC current measurement
⚠	Important safety information	A⎓	DC current measurement
⚡	Potentially dangerous voltage	Ω	Resistance measurement
⏚	Grounding	Hz	Frequency measurement
▣	Double insulation protection(2)	h_{FE}	Transistor measurement
⎓▭	Fuse	F	Capacitance measurement
🔋	Battery	℃	Temperature measurement
CE	Meet EU standards	⇥	Diode measurement
(MC)	China manufacturing measuring instrument license	•)))	On off measurement

Figure 1-9　Symbol description of digital multimeter

1.2.4.1　Voltage Measurement

(1) DC voltage measurement.

First, insert the black probe into the 'com' hole and the red probe into the 'V Ω'. Select the knob to the range larger than the estimated value (Note: the values on the dial are the maximum range, 'V-' represents DC voltage range, 'V~' represents AC voltage range, 'A' represents current range), then connect the probe to the power supply or both ends of the battery; keep the contact stable. The value can be read directly from the display screen. If the display is '1', it indicates that the measuring range is too small, so it is necessary to add a lot of range before measuring industrial electrical appliances. If '-' appears on the left side of the value, it indicates that the polarity of the probe is opposite to that of the actual power supply, and the red probe is connected to the negative electrode.

The measuring DC voltage with DMM is shown in Figure 1-10.

Figure 1-10　Measuring DC voltage with DMM

(2) AC voltage measurement.

The probe socket is the same as the DC voltage measurement, but the knob should be set to the range required by the AC Gear 'V ~'. There is no difference between positive and negative AC voltage, and the measurement method is the same as the previous one. Whether measuring AC or DC voltage, pay attention to personal safety, do not touch the metal part of the probe with your hand.

The measuring AC voltage with DMM is shown in Figure 1-11.

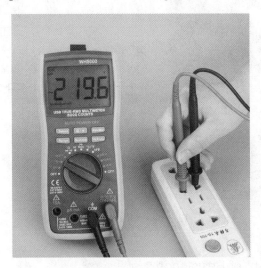

Figure 1-11 Measuring AC voltage with DMM

1.2.4.2 Current Measurement

(1) DC current measurement.

First insert the black probe into the 'COM' hole. If the current is greater than 200mA, insert the red probe into the '20A' Jack and turn the knob to the DC '20A'; if the current is less than 200mA, insert the red probe into the 'Ma' Jack and turn the knob to the appropriate range within 200mA DC. After adjustment, it can be measured. Put the multimeter in series into the circuit, keep stable, and then read. If '1' is displayed, the measuring range shall be increased; if '-' appears on the left side of the value, it indicates that the current flows from the black probe into the multimeter. Socket 'A' is protected by 200mA fuse, which will fuse when overload occurs, and shall be replaced in time according to the specified value. Jack '20A' has no fuse protection, the maximum current that can be continuously measured is 10a, and the measurement time shall be less than 15s.

(2) AC current measurement.

The measurement method is the same as that of DC current, but the gear should be set to AC gear. After the current measurement is completed, insert the red pen into the 'V Ω' hole. If you forget this step and directly measure the voltage, the multimeter and equipment will be damaged.

The measuring current with DMM is shown in Figure 1-12.

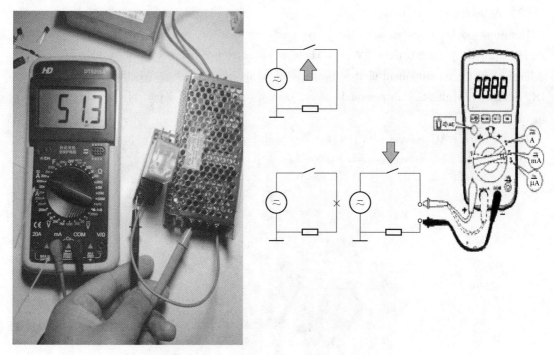

Figure 1-12 Measuring current with DMM

1.2.4.3 Measurement of Resistance

Insert the probe into the 'COM' and 'V Ω' holes, turn the knob to the required range in 'Ω', and connect the probe to the metal parts at both ends of the resistance. During the measurement, you can touch the resistance by hand, but do not touch the two ends of the resistance at the same time, which will affect the measurement accuracy - the human body is a conductor with large resistance but limited large. When reading, keep good contact between the probe and the resistance; note that the unit is 'Ω' in '200', kΩ in '2k' to '200k', and 'mΩ' in '2M' and above.

It should be noted that the polarity of the red probe is '+'; the open circuit is displayed as over range state, i.e. '1'; when measuring the online resistance, it is necessary to confirm that the power of the circuit under test has been turned off and the capacitance has been discharged before the measurement.

The Measuring resistance with DMM is shown in Figure 1-13.

1.2.4.4 Diode Measurement

Place the range switch in the diode symbol position. Insert the black probe into the 'COM' Jack, and the red probe into the 'V Ω' Jack. Connect the probe to the diode to be tested and it will display the voltage value of forward voltage drop. When the diode is reversed, it will display the over range state. When the input terminal is open, it will also display the over range state, that is, the highest position will display '1'. Test conditions: Forward DC current is about 1mA, re-

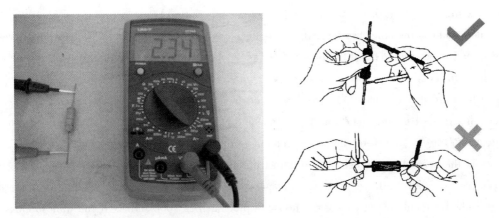

Figure 1-13 Measuring resistance with DMM

verse DC voltage is about 3V.

The reverse digital multimeter diode is shown in Figure 1-14.

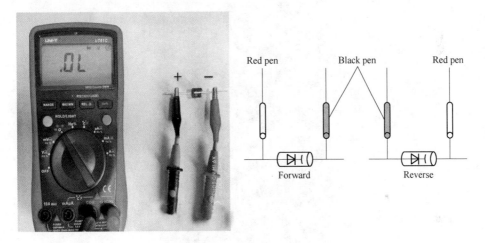

Figure 1-14 Reverse digital multimeter diode

1.2.4.5 Buzzer On-off Test

Insert the black probe into the 'COM' Jack, and the red probe into the 'V Ω' Jack. Place the range switch in the buzzer position. When the resistance between the two points is less than about 70Ω, the buzzer will sound.

Note: when the input end is open, the instrument displays over range status. The circuit to be tested must be checked on and off when the power supply is cut off, because any load signal may make the buzzer sound, leading to wrong judgment.

1.2.4.6 Measurement of Triode

The insertion position of the probe is the same as that of the diode. First, assume that the A-pin is the base, connect it with the black pen, and the red pen contacts the other two feet respectively;

if the readings are about 0.7V for both times, then connect the A-pin with the red pen, and the black pen contacts the other two feet, if '1' is displayed, then the A-pin is the base, otherwise, it needs to be remeasured, and the pipe is PNP pipe. How to judge collector and emitter? We can use 'h_{FE}' to judge: first, turn the gear to 'h_{FE}' and you can see a row of small jacks beside the gear, which are divided into PNP and NPN tube measurements. The pipe type has been determined. Insert the base electrode into the corresponding pipe type 'b' hole, and insert the other two feet into the 'c' and 'e' holes respectively. At this time, you can read the value, i.e. β value; fix the base electrode again, and adjust the other two feet; compare the two readings, and the pin position with the larger reading corresponds to the surface 'c' and 'e'. In addition: the upper method can only directly measure the small pipe such as 9000 series. If you want to measure the large pipe, you can use the wiring method, that is, use the small wire to lead out the three pins.

The schematic diagram of h_{FE} Jack measured by DMM is shown in Figure 1-15.

Figure 1-15　Schematic diagram of h_{FE} Jack measured by DMM

1.2.4.7　Measurement of MOSFET

N-channel has 3D01, 4D01 madein China, and Nissan's 3SK series. Determination of G-pole (grid): use diode gear of multimeter. If the positive and negative voltage drop between one pin and the other two pins is greater than 2V, it will display '1', which is grid G. Then exchange the probe to measure the other two legs. In the case of low voltage drop, the black probe is connected to the D pole (drain pole) and the red probe is connected to the S pole (source pole).

1.2.4.8　Precautions for Use of DMM

(1) If no number is displayed after power on, check whether the 9V integrated battery has

failed, check whether the battery lead is broken and whether the battery clamp contacts firmly. If the low voltage sign is displayed, replace the battery in time. When measuring, if only the highest digit '1' is displayed, and other bits are eliminated, it is proved that the instrument has been overloaded, a higher range should be selected. Some digital multimeter with reading holding switch or key should be placed in off position at ordinary times to avoid affecting normal measurement. Some new digital multimeters add the function of automatic shutdown. When the meter stops using or stops in a certain gear for more than 15 minutes, it can automatically cut off the power supply and make the meter in the 'sleep' state of low power consumption, instead of failure. At this time, you only need to restart to return to normal operation.

(2) When using the digital multimeter, the specified limit value shall not be exceeded. The input voltage limit value of the highest DC V gear is 1000V, and that of the highest AC V gear is 700V or 750V (effective value). When DC voltage is superimposed on the measured AC voltage, the sum of the two voltages shall not exceed the limit value of the input voltage of the used AC V gear.

(3) When measuring alternating current, the black probe shall be connected to the low potential terminal of the measured voltage (such as the common ground of the measured signal source, the case, the neutral terminal of 220V alternating current, etc.), so as to eliminate the influence of the instrument input terminal on the ground distribution capacitance and reduce the measurement error.

(4) When measuring DC voltage (or DC current), the instrument can automatically determine and display the polarity of voltage (or current), so it is unnecessary to consider the connection method of the probe.

(5) When measuring the high current, a '20A' Jack shall be used. The jack is not equipped with a protection device, so the time for measuring the high current shall not exceed 10~15s, otherwise the accuracy of the reading will be affected.

(6) The red lead of the digital multimeter is positively charged and the black lead is negatively charged, which is exactly the opposite of the polarity of the resistance block of the pointer multimeter. When measuring the components with polarity, pay attention to the polarity of the probe.

(7) When measuring the resistance, do not touch the metal end of the probe or the leading out end of the element with both hands, so as not to introduce the resistance into the body and affect the measurement result. It is forbidden to measure the resistance when the line under test is live.

(8) When measuring electrolytic capacitor with capacitor block, the polarity of the measured capacitor shall be consistent with that indicated by the capacitor socket. Before measurement, the capacitor must be discharged to avoid damaging the instrument.

1.2.5 Bill of Materials

The table of the bill of materials is shown in Table 1-4.

Table 1-4 Bill of materials

No.	Name	Quantity
1	XK-SX2C advanced maintenance electrician training platform	1
2	One maintenance electrician training component (XKDT11)	1
3	One maintenance electrician training component (XKDT12A)	1
4	Digital multimeter	1
5	Electronic components Kit	1

1.2.6 Task Implementation

The table of the mission statement is shown in Table 1-5.

Table 1-5 Mission statement

( ) mission statement				
1. According to the requirements, complete the required devices and complete the following table				
No.	Device name	Number of devices	Device model	Whether the device is in good condition	Remarks
2. Project task description				

Continued Table 1-5

( ) mission statement
3. Introduction to the working principle involved in the task
4. Operation process record of measuring a component with DMM
5. Record of measurement results
6. Task summary

1.2.7 Task Evaluation

The table of the scoring table is shown in Table 1-6.

Table 1-6 Scoring table

	(	) scoring table			

Device inventory and measurement (10 points)

No.	Key inspection contents	Scoring criteria	Weighting	Score	Remarks
1	Function verification of digital multimeter	Deduct 1 point for checking errors	5		
2	Device inventory	Deduct 1 point for each error in device counting	5		
		Subtotal			

Measurement of current and voltage (20 points)

No.	Key inspection contents	Scoring criteria	Weighting	Score	Remarks
1	Current measurement	Deduct 1 point for one mistake	10		
2	Voltage measurement	Deduct 1 point for one mistake	10		
		Subtotal			

Working principle description of component measurement (10 points)

No.	Key inspection contents	Scoring criteria	Weighting	Score	Remarks
1	Characteristics of components	Is it complete and discretionary	5		
2	Working principle	Is it complete and discretionary	5		
		Subtotal			

Steps of measuring components with DMM (30 points)

No.	Key inspection contents	Scoring criteria	Weighting	Score	Remarks
1	Lead wiring	Deduct 1 point for one mistake	10		
2	Gear selection	Deduct 1 point for one mistake	10		
3	Measuring position of components	Deduct 1 point for one mistake	10		
		Subtotal			

Measurement results (20 points)

No.	Key inspection contents	Scoring criteria	Weighting	Score	Remarks
1	Whether the measurement data is accurate	Understand or not	10		
2	Component damage determination	Understand or not	10		
		Subtotal			

Continued Table 1-6

( ) scoring table						
Professional quality (10 points)					
No.	Key inspection contents	Scoring criteria	Deduction	Score	Remarks
1	Live operation	Deduct 5 points at a time until all points are deducted			
2	Standard operation	Deduct 1 point at a time until all points are deducted			
3	Team work	As appropriate			
4	Clean working position	As appropriate			
		Total			

1.2.8 Task Development

Common faults and maintenance of digital multimeter

(1) Common faults. To find the fault, we should first look outside and then inside, first easy and then difficult, break it up into parts, and make a breakthrough. The methods can be roughly divided into the following:

1) The sensory method makes a direct judgment on the cause of the fault by means of the senses. Through the visual inspection, it can be found that the causes of the abnormal temperature rise can be found, such as broken wire, desoldering, short circuit of grounding wire, broken fuse tube, burned element, mechanical damage, copper foil warping and fracture on the printed circuit, etc.; the temperature rise of the battery, resistance, crystal tube and integrated block can be touched, and the cause of the abnormal temperature rise can be found by referring to the circuit diagram. In addition, it can also check whether the components are loose, whether the integrated circuit pin is firmly inserted, whether the transfer switch is jammed; it can hear and smell whether there is abnormal sound and peculiar smell.

2) The method of measuring voltage can find out the fault point quickly. For example, measure the working voltage and reference voltage of A-D converter.

3) Short circuit method is generally used in the inspection of A-D converter mentioned above. This method is widely used in the repair of weak current and micro electric instruments.

4) The open circuit method interrupts the suspicious part from the whole machine or unit circuit. If the fault disappears, it means that the fault is in the disconnected circuit. This method is mainly suitable for short circuit.

5) When the fault has been reduced to a certain place or several components, it can be measured online or offline. If necessary, replace it with a good component. If the fault disappears, the component is broken.

6) The interference method uses the induced voltage of human body as the interference signal to observe the change of liquid crystal display. It is often used to check whether the input circuit and

display part are intact.

(2) Troubleshooting. For a fault instrument, first check and judge whether the fault phenomenon is common (all functions cannot be measured), or individual (individual function or individual range), then distinguish the situation and solve the problem.

If all gears fail to work, check the power circuit and A-D converter circuit. When checking the power supply part, remove the laminated battery, press the power switch, connect the positive probe to the negative power supply of the meter under test, and connect the negative probe to the positive power supply (for digital multimeter). Turn the switch to the secondary tube measurement file. If the positive voltage of the secondary tube is displayed, it means that the power supply part is good. If the deviation is large, it means that there is a problem in the power supply part. In case of open circuit, check the power switch and battery lead. If there is a short circuit, it is necessary to use the open circuit method to gradually disconnect the components using the power supply, focusing on checking the operational amplifier, timer, A-D converter, etc. If there is a short circuit, generally more than one integrated component is damaged. Check that the A-D converter can be carried out at the same time as the basic meter, which is equivalent to the DC meter of the analog multimeter. The specific inspection method is: first, turn the range of the tested meter to the lowest DC voltage; second, measure whether the working voltage of the A-D converter is normal. According to the model of A-D converter used in the table, corresponding to V_+ pin and COM pin, whether the measured value is consistent with its typical value. Thirdly, measure the reference voltage of A-D converter. At present, the reference voltage of commonly used digital multimeter is generally 100mV or 1V, that is, measure the DC voltage between U_{REF+} and COM. If it deviates from 100mV or 1V, it can be adjusted by external potentiometer. Check the display number whose input is zero, short-circuit the positive terminal IN_+ and negative terminal IN_- of the A-D converter so that the input voltage $U_{in} = 0$, and the instrument displays '00.0' or '00.00'. Check the display for full bright strokes. Connect the TEST pin of the test terminal and the V_+ of the positive power supply terminal to make the logic ground become high potential and all digital circuits stop working. Since DC voltage is applied to each stroke, the bright alignment table of all strokes displays '1888', and the alignment table displays '18888'. If there is a lack of stroke, check whether there is poor contact and disconnection between the corresponding output pin of A-D converter and the conductive adhesive (or connecting wire) and the display.

If there is a problem with some files, the A-D converter and the power supply are working normally. Because DC voltage and resistance share a set of voltage dividing resistor; AC and DC current share a shunt; AC voltage and AC current share a set of AC-DC converter; others, such as Cx, h_{FE}, F, etc., are composed of independent different converters. Understand the relationship between them, and then according to the power diagram, it is easy to find the fault location. If the measurement small signal is not accurate or the display digital jitter is large, it is important to check whether the contact of the range switch is good.

If the measurement data is unstable, and the value always increases accumulatively, short circuit the input of A-D converter, and the display data is not zero, it is generally caused by the poor performance of 0.1μF reference capacitance.

According to the above analysis, the basic repair sequence of digital multimeter should be: digital meter head → DC voltage → DC current → AC voltage → AC current → resistance gear (including buzzer and check the positive voltage drop of secondary pipe) → Cx → h_{FE}, F, H, T, etc. But we should not be too mechanical. We can deal with some obvious problems first. However, during the adjustment, the above procedures must be followed.

In short, a fault multimeter, after proper detection, first of all, it is necessary to analyze the possible parts of the fault, and then find the fault location according to the circuit diagram for replacement and repair. Because the digital multimeter is a more precise instrument, it is necessary to replace the components with the same parameters, especially the A-D converter. It is necessary to use the integration block strictly selected by the manufacturer, otherwise the required accuracy will not be achieved due to the error. The new A-D converter also needs to be checked according to the method mentioned above, and it must not be trusted for being new.

At present, there are many domestic manufacturers of digital multimeter, and the quality is good or bad. It is not easy to find the quality problems of double-sided copper clad plate in the repair. When the insulation strength of resin board is not enough, the main performance is that the error is large when measuring the high voltage, so it should be distinguished from the change of resistance value of the partial voltage resistance when repairing. In this case, it is better to use the open circuit method to find the fault point. The burned and carbonized parts shall be cleaned to meet the insulation requirements. When the double-sided connection fails to input signal due to the fracture of the transition hole, it is easy to be confused with the phenomenon of poor transfer switch and difficult to separate. For this kind of fault, the short circuit method should be used to find the fault point.

The table of the troubleshooting record is shown in Table 1-7.

Table 1-7 Troubleshooting record

		Troubleshooting record
Fault 1	Fault phenomenon	
	Cause of fault	
	Troubleshooting process	
Fault 2	Fault phenomenon	
	Cause of fault	
	Troubleshooting process	

Continued Table 1-7

		Troubleshooting record
Fault 3	Fault phenomenon	
	Cause of fault	
	Troubleshooting process	
Fault 4	Fault phenomenon	
	Cause of fault	
	Troubleshooting process	

1.2.9 Task Summary

When measuring the resistance of the digital multimeter, no matter whether the tested object is a complex system or a single electronic component, there shall be no current component formed by any power source other than the test current of the digital multimeter in the tested part, and no part of the body such as the finger of the measurer shall contact the contact pin of two probes and the conductive part of the tested object. In the process of measuring high voltage and current, do not switch the gear of the selector switch, otherwise it is easy to burn the switch contacts and damage other electronic components inside the instrument. If it is necessary to switch due to special circumstances, the probe shall be removed from the circuit to be measured before switching. When the digital multimeter is in the diode test gear, buzzer gear and resistance gear, the red probe is connected to the high potential inside the instrument and has positive electricity, while the black probe is connected to the virtual ground and has negative electricity.

The AC voltage block of ordinary digital multimeter belongs to the average value instrument, and it is designed according to the sine wave characteristics, so it can not be directly used to measure the non sine wave voltage such as sawtooth wave, triangle wave and rectangle wave. Even if the sine wave voltage is measured, if the waveform distortion is large, the correct measurement results will not be obtained. We should pay attention to all these problems when we measure.

Project 2 Basic Control Skill Training of Three-phase Asynchronous Motors

Task 2.1 Cognition and Basic Test of Three-phase Squirrel-cage Asynchronous Motor

2.1.1 Task Description

Xiao Wang, one of the front-line employees in a factory, accidentally touched the shell of the running motor on the back of his hand during his work. His body felt numb, so he quickly withdrew his hand. After that, Xiao Wang shut down the motor, cut off the power of the equipment, and tested and repaired the motor.

Let's think about it. Why does Xiao Wang feel numb? Is it normal for the body to feel numb when touching the motor?

This task aims at the above problems, learning the structure and working principle of the three-phase squirrel cage asynchronous motor, as well as the performance of the motor through the nameplate.

In addition, in actual use, it is necessary to ensure that the insulation between the motor shell and winding is good, so as to avoid safety problems. How to test it?

2.1.2 Task Target

(1) Familiar with the structure of three-phase squirrel cage asynchronous motor.

(2) Master the working principle of three-phase squirrel cage asynchronous motor.

(3) Palm can understand the nameplate of three-phase squirrel cage asynchronous motor.

(4) Master the method of testing the insulation of three-phase squirrel cage asynchronous motor.

(5) Master the disassembly and installation of three-phase squirrel cage asynchronous motor.

(6) Master safety operation precautions.

2.1.3 Task Analysis

In the task description, Xiao Wang, an employee, accidentally touches the running motor and feels numb. The reason is that the insulation of the motor shell is poor, and the motor shell with good performance has good insulation and cannot be charged. This requires us to master the method of detecting the insulation of motor shell, master the knowledge and skills of motor structure, working principle, etc., so as to disassemble and repair the motor.

In production practice, motor is the prime mover of production machinery, whose function is to convert electric energy into mechanical energy. Machine tools, cranes, forging machines, blowers, pumps, most of the production machinery are driven by it. According to the different rotor structure, the motor can be divided into squirrel cage motor and winding motor. The three-phase squirrel cage asynchronous motor has the advantages of simple structure, reliable operation, low price, convenient use, installation and maintenance, and is widely used in various fields. The outline of three-phase asynchronous motor is shown in Figure 2-1.

Figure 2-1 Configuration of three-phase squirrel-cage asynchronous motor

This task is divided into four sub tasks:
(1) The structure of three-phase squirrel cage asynchronous motor;
(2) Working principle of three-phase asynchronous motor;
(3) Nameplate data of three-phase asynchronous motor;
(4) Inspection of three-phase squirrel cage asynchronous motor.

2.1.4 Task-related Knowledge

2.1.4.1 The Structure of Three-phase Squirrel-cage Asynchronous Motor

(1) Stator part. The stator of three-phase asynchronous motor produces rotating magnetic field. The stator is mainly composed of frame, stator core, stator winding, etc. The stator core is a part of the magnetic circuit of the motor. It is made of silicon steel sheets which are insulated from each other and stacked into a cylinder, which is installed on the inner wall of the frame. Slot holes are punched evenly on the inner circumference surface of the stator core to embed the stator winding. The stator core is shown in Figure 2-2, and the stator winding is shown in Figure 2-3.

Figure 2-2 Stator core   Figure 2-3 The stator winding

(2) Rotor part. The rotor action of three-phase asynchronous motor is to produce induced electromotive force or current under the action of stator magnetic field and to rotate and output mechanical energy. The rotor is mainly composed of shaft, core and winding. The shaft is pressed with a cylindrical rotor core made of silicon steel sheets. The rotor core is also part of the magnetic circuit of the motor. Slot holes are evenly punched on the outer circumference surface of the rotor core, and the rotor winding is embedded in the slot. It can be divided into cage type and winding type. Cage rotor core and winding are shown in Figure 2-4, and winding rotor core and winding are shown in Figure 2-5.

Figure 2-4 Cage rotor core and winding   Figure 2-5 Core and winding of wound rotor

(3) Structure split of three phase asynchronous motor.

The structure split diagram of three-phase asynchronous motor is shown in Figure 2-6.

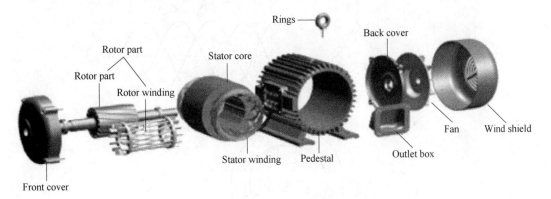

Figure 2-6 Structure split of three-phase asynchronous motor

2.1.4.2 Working Principle of Three-phase Asynchronous Motor

(1) Generation of rotating magnetic field. The three-phase symmetrical alternating current (shown in Figure 2-7) is connected to the three-phase stator winding (the three coils are 120° apart from each other and distributed on the circumference of the inner circle of the stator core). The stator winding of three-phase asynchronous motor is connected with three-phase symmetrical alternating current, as shown in Figure 2-8.

By analyzing the position change of the magnetic field produced in different time, it can be known that the rotating magnetic field will be produced when the three-phase alternating current is connected to the three-phase stator winding. The generation of rotating magnetic field is shown in Figure 2-9.

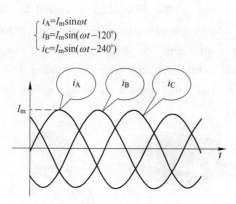
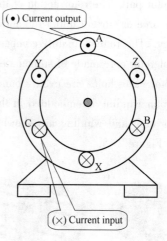

$\begin{cases} i_A = I_m \sin\omega t \\ i_B = I_m \sin(\omega t - 120°) \\ i_C = I_m \sin(\omega t - 240°) \end{cases}$

Figure 2-7 Three phase symmetrical alternating current

Figure 2-8 Stator winding of three-phase asynchronous motor is connected with three-phase symmetric alternating current

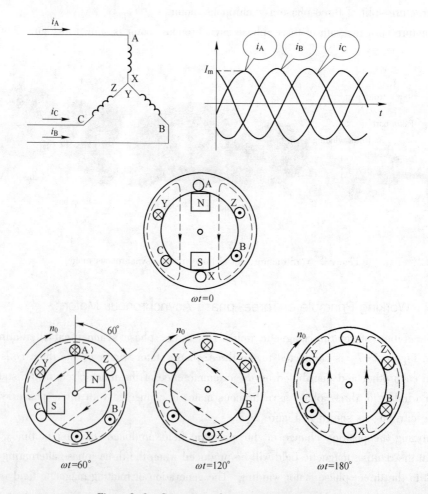

Figure 2-9 Generation of rotating magnetic field

(2) Principle of rotor rotation. The principle of rotor rotation is shown in Figure 2-10.

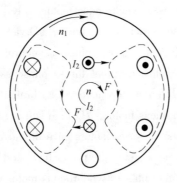

Figure 2-10   Principle of rotor rotation

Working principle of three-phase asynchronous motor: three-phase alternating current is connected into the stator winding of three-phase asynchronous motor to generate rotating magnetic field. The static rotor has a relative motion of cutting magnetic field line with respect to rotating magnetic field to generate induced electromotive force and induced current. There is current on the rotor winding, which will be affected by electromagnetic force in the magnetic field, forming electromagnetic torque T, overcoming the resistance torque drives the rotor to rotate and realizes the purpose of converting electrical energy into mechanical energy.

(3) Direction of rotation of rotating magnetic field. The direction of rotation of the rotating magnetic field depends on the phase sequence of the three-phase current. Changing the current phase sequence flowing into the three-phase winding can change the rotation direction of the rotating magnetic field; changing the rotation direction of the rotating magnetic field can also change the rotation direction of the three-phase asynchronous motor. The influence of phase sequence of three-phase current on the steering of three-phase asynchronous motor is shown in Figure 2-11.

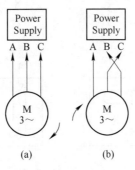

Figure 2-11   Influence of phase sequence of three-phase current on three-phase asynchronous motor steering
(a) Forward; (b) Reverse

2.1.4.3   Nameplate Data of Three-phase Asynchronous Motor

(1) Model. Model is the code of different kinds and types of motors, each letter of which has a

certain meaning. Commonly used Y series asynchronous motors include Y (IP44) closed type, Y (IP23) protective small three-phase asynchronous motor, YR (IP44) closed type, YR (IP23) protective wound type three-phase asynchronous motor, YD variable pole multi speed three-phase asynchronous motor, YX high efficiency three-phase asynchronous motor, YH high slip three-phase asynchronous motor, YB explosion-proof three-phase asynchronous motor, YCT electromagnetic speed regulation three different step motor, YEJ brake three-phase asynchronous motor, YTD elevator three-phase asynchronous motor, YQ high starting torque three-phase asynchronous motor and dozens of other products. Y series motors have the advantages of high efficiency, energy saving, good characteristics, low noise, etc. the power level and installation size meet the international standards. The model of three-phase asynchronous motor is shown in Y132M-4.

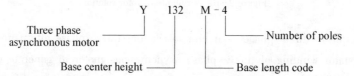

(2) Rated capacity (power). Refers to the output power of the motor shaft when the motor is in rated operation.

(3) Frequency. It refers to the allowable frequency added to the stator winding of the motor.

(4) Rated voltage. The line voltage to be applied to the stator winding under the specified connection method.

(5) Rated current. The line current of the stator winding under the specified connection method.

(6) Connection mode. There are two connection methods of three-phase load: star and corner. Two connection methods of stator winding of three-phase asynchronous motor are shown in Figure 2-12.

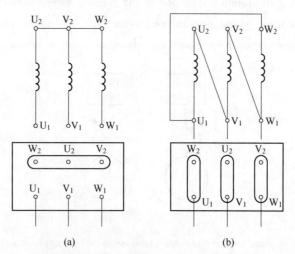

Figure 2-12 Two connection methods of stator winding of three-phase asynchronous motor
(a) Star joint; (b) Corner joint

(7) Speed. Speed on motor shaft (n).

(8) Insulation class. The insulation grade is determined by the insulating materials used for the

motor. According to the heat resistance degree, the insulating materials are divided into A, E, B, F, H and other grades.

(9) How it works. According to the continuous operation time, the motor is designed to work in three ways: continuous (S_1), short-time (S_2), repeated short-time (S_3).

Continuous operation means that the motor can operate continuously for a long time according to the power specified on the nameplate.

Short time operation means that the motor can not operate continuously. When the rated power is output, it can only operate for a short time as specified on the nameplate.

Repetitive short-time operation means that the motor can not operate continuously at rated power output, and can only operate repeatedly for a specified time.

(10) Power factor ($\cos\varphi 1$). Refers to the power factor of the motor at rated operation.

(11) Efficiency. It refers to the ratio of motor output power to input power.

2.1.4.4 Inspection of Three-phase Squirrel Cage Asynchronous Motor

Necessary inspection shall be made before using the motor.

(1) Mechanical inspection. Check whether the outgoing line is complete and reliable, whether the rotor rotates flexibly and evenly, and whether there is abnormal sound, etc.

(2) Electrical inspection. In the daily operation of the motor, the coil is often loose, which makes the insulation wear and aging, or the surface is polluted, damped and so on, which causes the insulation resistance to decline day by day. The reduction of the insulation resistance to a certain value will affect the starting and normal operation of the motor, and even damage the motor, endangering personal safety. Therefore, the insulation resistance of each phase winding to the casing and the insulation resistance between windings shall be measured before the use of various motors or after the mould season, damp and reinstallation. The measurement of insulation resistance is generally carried out with a megger. Learning how to use the megger can bring us convenience when checking the insulation of motors, electrical appliances and lines and measuring high value resistance.

Megger is a kind of simple and commonly used instrument for measuring high resistance. It is mainly used to check the insulation resistance of electrical equipment, household appliances or electrical lines to the ground and between phases, so as to ensure that these equipment, appliances and lines work in a normal state and avoid accidents such as electric shock casualties and equipment damage. Megger is mostly powered by hand motor, its scale is in megaohms ($M\Omega$).

The common megohmmeter is mainly composed of two parts: high-voltage manual motor and magnetoelectric current ratio meter. The structure diagram and schematic circuit diagram of the megohmmeter are shown in Figure 2-13.

1) Working principle of megameter. There are two coils connected with the probe of the megger, one is connected in series with the additional resistance R in the meter, the other is connected in series with the measured resistance R_X, and then connected to the hand-held generator together. When the motor is shaken by hand, there is current passing through the two coils at the same

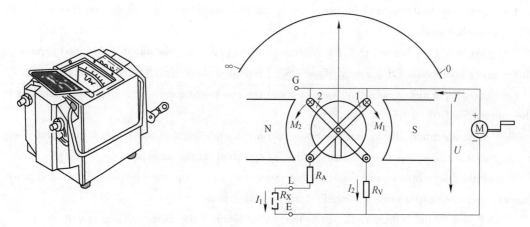

Figure 2-13 Appearance and circuit diagram of megger

time, which produces torque in opposite directions on the two coils. The needle deflects an angle with the composite torque of the two torques. The deflection angle depends on the ratio of the two currents, and the additional resistance is constant, so the current value only depends on the size of the resistance to be measured.

2) Selection of megohmmeter. Before measuring the insulation resistance of electrical equipment, the appropriate megger shall be selected according to the nature and voltage level of the equipment to be tested.

Generally, when measuring equipment with rated voltage below 500V, use a megger of 500~1000V, and when measuring equipment with rated voltage above 500V, use a megger of 1000~2500V. For example, when measuring the insulation resistance of high-voltage equipment, the megger with rated voltage below 500V cannot be used, because the measurement results cannot reflect the insulation resistance under the working voltage; similarly, the megger with too high voltage cannot be used to measure the insulation resistance of low-voltage electrical equipment, otherwise the insulation of the equipment will be damaged.

In addition, the measuring range of the megger shall also be consistent with the range of the insulation resistance to be measured. Generally, attention shall be paid not to make the measurement range exceed the insulation resistance value to be measured too much, so as to avoid large error in reading. Generally, when measuring the insulation resistance of low-voltage electrical equipment, $0 \sim 200 M\Omega$ range meter can be selected. When measuring high-voltage electrical equipment or cables, $0 \sim 2000 M\Omega$ range meter can be selected. The scale is not from zero, but from $1 M\Omega$. Generally speaking, megohmmeter is not suitable for measuring the insulation resistance of low-voltage electrical equipment.

3) Inspection before use. Before using the megger, an open circuit and short circuit test shall be carried out to check whether the megger is in good condition.

Open the 'L' and 'E' ends, shake the handle, the pointer should be at '∞', and then short the 'L' and 'E' ends, shake the handle, the pointer should be at '0', indicating that the megger is good, otherwise there is error.

Megohmmeter open circuit and short circuit test are shown in Figure 2-14.

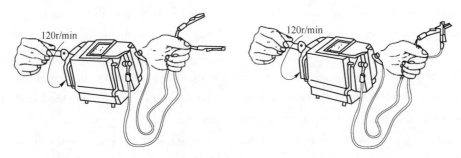

Figure 2-14 Megohmmeter open circuit and short circuit test

4) Wiring of megger. There are three terminals marked with ground 'E', circuit 'L' and protection ring 'G' on the megger.

① When measuring the insulation resistance of the circuit, the tested terminal can be connected to the 'L' terminal of the circuit, and a good ground wire can be connected to the terminal of the grounding 'E' as shown in Figure 2-15 (a).

② When measuring the insulation resistance of the motor, connect the motor winding to the 'L' end of the circuit and the housing to the 'E' end of the grounding, as shown in Figure 2-15 (b).

③ Measure the insulation performance between the motor windings, and connect the 'L' end and the 'E' end of the circuit to the terminals of the two windings of the motor.

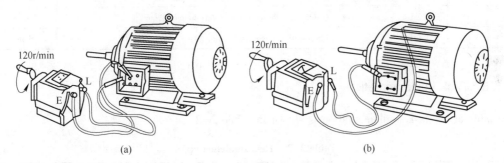

Figure 2-15 use megger to check the insulation performance between motor windings
and between windings and casings

(a) Insulation inspection between motor windings; (b) Insulation inspection between motor shell and monthly windings

④ When measuring the insulation resistance of the cable core to the cable shell, in addition to connecting the cable core to the circuit 'L' end and the cable shell to the grounding 'E' end, the inner insulation between the cable shell and the core shall be connected to the protection ring 'G' to eliminate the error caused by the surface leakage.

5) Precautions for using megger:

① Before the measurement, the power supply shall be cut off first, and the tested equipment must be discharged (about 2~3min), so as to ensure the safety of equipment and personnel.

② The wires connected between the terminal and the tested equipment shall not be double insulated wire or stranded wire. Single stranded wire shall be used for separate connection to avoid error caused by poor insulation of stranded wire. The equipment surface shall be kept clean and dry.

③ During measurement, the surface shall be placed stably, and the handle shall be shaken from slow to fast.

④ Generally, the pointer position after shaking evenly for 1min is taken as the reading. Generally 120r/min. If the pointer indicates 0 during the measurement, stop turning the handle to prevent the coil in the meter from overheating and burning out.

⑤ When the rotation of the megger is not stopped or the tested equipment is not discharged, it is not allowed to remove the wire by hand to avoid electric shock.

2.1.5 Bill of Materials

The table of the bill of materials is shown in Table 2-1.

Table 2-1 Bill of materials

Serial number	Name	Number
1	Xk-SX2C advanced maintenance electrician training platform	1 station
2	Three phase asynchronous motor	1 station
3	Jumper	Some
4	ZC25-4 type megger	Some
5	Multimeter	Some
6	Work clothes, insulating shoes, safety helmets, etc	Some
7	Screwdriver, wrench, casing, vise	Some

2.1.6 Task Implementation

The table of the task implementation is shown in Table 2-2.

Table 2-2 Task implementation

Understanding and basic test assignment of three-phase squirrel cage asynchronous motor					
1. According to the requirements, complete the required device inventory and complete the following table					
Serial number	Device name	Number of devices	Device parameters	Device measurement	Remarks

Continued Table 2-2

Understanding and basic test assignment of three-phase squirrel cage asynchronous motor

1. According to the requirements, complete the required device inventory and complete the following table

Serial number	Device name	Number of devices	Device parameters	Device measurement	Remarks

2. Description of working principle of motor

3. Motor disassembly process record

4. Motor insulation performance test process record

5. Task summary

2.1.7 Task Evaluation

The table of the task evaluation is shown in Table 2-3.

Table 2-3 Task evaluation

Scoring table for motor disassembly and inspection

Device inventory and measurement (10 points)

Serial Number	Key inspection contents	Scoring criteria	Score	Score	Remarks
1	Device inventory	Deduct 1 point for each error	5		
2	Instrument use, gear selection	Deduct 1 point for one item with wrong use and gear	5		
		Subtotal			

Motor disassembly (30 points)

Serial Number	Key inspection contents	Scoring criteria	Score	Score	Remarks
1	Disassembly is correct and placed in orde	Deduct 2 points for one mistake	10		
2	Undamaged motor components	Deduct 10 points for one damage	20		
		Subtotal			

Motor insulation test (20 points)

Serial Number	Key inspection contents	Scoring criteria	Score	Score	Remarks
1	The megger is set correctly before use	Deduct 5 points for one mistake	10		
2	Correct use of megger	Deduct 5 points for one mistake	10		
		Subtotal			

Motor assembly (30 points)

Serial Number	Key inspection contents	Scoring criteria	Score	Score	Remarks
1	Is the installation sequence correct	Deduct 5 points for one mistake	10		
2	Whether the device is damaged	Deduct 5 points for one mistake	10		
3	Whether the assembly is firm	Deduct 5 points for one mistake	10		
		Subtotal			

Professionalism (10 points)

Serial Number	Key inspection contents	Scoring criteria	Score	Score	Remarks
1	Live operation	Deduct 5 points at a time until all points are deducted			
2	Standard operation	Deduct 1 point at a time until all points are deducted			
3	team work	As appropriate			
4	Clean working position	As appropriate			
		Total			

2.1.8 Troubleshooting

(1) Motor case is live. Generally, leakage current of motor shall not be greater than 0.8mA to ensure personal safety.

The main causes of the leakage of the motor case are that the insulation of an outgoing line in the motor is damaged and touches the case; the local burning of the motor winding causes the leakage between the stator and the case. What is more common is that the motor is in a high humidity environment for a long time, which leads to the reduction of the damp insulation of the motor and the electrification of the casing. At this time, use a megger to measure the insulation resistance between each winding and the casing of the motor. If it is less than $2M\Omega$, it means that the motor has been seriously damped, and the stator winding of the motor should be baked to remove the moisture.

(2) Stator and rotor core troubleshooting. The stator and rotor are made of silicon steel sheets which are insulated from each other. They are the magnetic circuit of the motor. The damage and deformation of stator and rotor core are mainly caused by the following reasons.

1) Excessive wear or poor assembly of bearings will cause friction between stator and rotor, damage the surface of iron core, and then cause short circuit between silicon steel sheets, increase the iron loss of motor, and make the temperature rise of motor too high. At this time, the burr shall be removed with fine file and other tools, and the short circuit of silicon steel sheet shall be eliminated. After cleaning, the insulating paint shall be applied, and then it shall be heated and dried.

2) Remove the old winding with too much force, so that the slot is tilted and opened outwards. At this time, pointed nose pliers, wooden hammers and other tools shall be used to repair, so as to reset the teeth, and hard insulating materials such as green shell paper and rubber board shall be added between the silicon steel sheets with gaps that are not easy to reset.

3) The iron core surface is rusted due to moisture and other reasons. At this time, sandpaper shall be used to polish it, and insulating paint shall be applied after cleaning.

4) The high heat generated by grounding around the group will burn the iron core or teeth. Use chisel or scraper and other tools to remove the deposit and apply insulating paint to dry.

5) The connection between the iron core and the base is loose, and the original set screw can be tightened. If the set screw fails, the set hole can be re drilled and tapped on the base, and the set screw can be tightened.

(3) Bearing troubleshooting. The shaft is supported by the bearing to rotate, which is the heaviest part of the load, and is easy to wear.

1) Fault check.

① In serviceinspection: When the rolling bearing is short of oil, it will hear the sound of 'Gu Lu Gu Lu'; if the sound of discontinuous 'stem' is heard, it may be that the bearing steel ring is broken. There will be slight noise when sand and other sundries are mixed in the bearing or the bearing parts are slightly worn.

② Inspection after disassembly: First, check whether the bearing rolling body and inner and outer steel rings are damaged, rusted, scarred, etc., then hold the inner ring of the bearing by hand, and make the bearing level, and push the outer rigid ring with the other hand. If the bearing is good, the outer steel ring shall rotate stably, without vibration and obvious stagnation during the rotation, and the outer steel ring shall not reverse after the rotation stops. Otherwise, the bearing can no longer be used.

The left hand catches the outer ring, the right hand holds the inner steel ring and pushes hard in all directions. If it feels loose when pushing, it is seriously worn.

2) Troubleshooting. The rust spots on the outer surface of the bearing can be wiped with 00 abrasive paper, and then put into the gasoline for cleaning; or when the bearing has cracks, the inner and outer rings are broken or the bearing is excessively worn, a new bearing should be replaced. When replacing a new bearing, select the same bearing as the original model.

(4) Shaft troubleshooting:

1) Shaft bending. If the bending is not large, it can be repaired by polishing the journal and slip ring; if the bending is more than 0.2mm, the rotating shaft can be placed under the press, pressurized and corrected at the beat bending position, and the corrected shaft surface can be cut and polished by lathe; if the bending is too large, it needs to be replaced with a new shaft.

2) Journal wear. When the journal wear is not large, a layer of chromium can be plated on the journal, and then ground to the required size; when there is more wear, hardfacing can be carried out on the journal, and then cutting and polishing can be carried out on the lathe; if the journal wear is too large, 2~3mm can also be cut on the journal, and then a sleeve can be turned, while it is hot, and then turned to the required size.

3) Shaft crack or fracture. When the transverse crack depth of the shaft is not more than 10%~15% of the shaft diameter and the longitudinal crack is not more than 10% of the shaft length, it can be remedied by the surfacing method, and then fine turning to the required size. If the crack of the shaft is serious, it is necessary to replace it with a new one.

(5) Maintenance of casing and end cover. If there is any crack in the shell and end cover, it shall be repaired by overlay welding. If the bearing boring clearance is too large, resulting in the loose fit of the bearing end cover, the bearing hole wall can be deburred evenly with a punch, and then the bearing can be driven into the end cover. For motors with higher power, the method of patching or electroplating can also be used to process the required size of bearings.

The table of the troubleshooting record is shown in Table 2-4.

Table 2-4　Troubleshooting record

		Troubleshooting record
Fault 1	Fault phenomenon	
	Cause of fault	
	Troubleshooting process	

Continued Table 2-4

		Troubleshooting record
Fault 2	Fault phenomenon	
	Cause of fault	
	Troubleshooting process	
Fault 3	Fault phenomenon	
	Cause of fault	
	Troubleshooting process	
Fault 4	Fault phenomenon	
	Cause of fault	
	Troubleshooting process	

2.1.9 Task Development

Disassembly and assembly of motor, See Figure 2-6 for the structure split diagram of three-phase asynchronous motor.

(1) General disassembly steps of three-phase asynchronous motor:

1) Cut off the power andremove the belt.
2) Remove the power wiring and ground wire from the junction box.
3) Remove the foot nut, spring washer and flat gasket.
4) Remove the pulley.
5) Remove the front bearing cover.
6) Remove the front end cover. A flat chisel of suitable size can be used to insert it into the protruding ear of the end cover, and pry outward in turn according to the diagonal of the end cover until the front cover is removed.
7) Remove the shroud and the vanes.
8) Remove the rear bearing cover.
9) Remove the rear end cover and rotor.

Before drawing out the rotor, a thick cardboard should be placed between the bottom of the rotor and the end of the stator winding to avoid damaging the stator core and winding when drawing out the rotor.

10) Finally, remove the front and rear bearings and bearing inner cover with puller.

(2) General assembly steps of three-phase asynchronous motor: The motor shall be assembled in reverse order of disassembly.

1) Installing the bearing.

2) Installing the rotor.

3) Install the end cover.

4) Install the bearing cover.

5) Installing fan and hood.

6) Installation of pulley or coupling.

7) General inspection after assembly.

(3) Several methods of bearing installation. Preparation before installation:

1) After cleaning the bearing and bearing cover with kerosene, check whether there is crack in the bearing and whether there is rust in the raceway.

2) Then rotate the outer ring of the bearing by hand, and observe whether it rotates flexibly and evenly. To determine whether the bearing should be replaced.

3) If it does not need to be replaced, wash the bearing with gasoline and dry it with a clean cloth for installation. When replacing a new bearing, it should be placed in 70~80℃ transformer oil, heated for about 5min, after all the antirust oil is dissolved, it should be cleaned with gasoline, dried with a clean cloth for installation.

Several common installation methods:

(1) Knock. Put the bearing on the shaft and align it with the journal. Use a section of iron pipe whose inner diameter is slightly larger than the journal diameter, and whose outer diameter is slightly larger than the outer diameter of the bearing inner race. Put one end of the iron pipe on the inner race of the bearing. Use a hammer to knock the other end of the iron pipe and knock the bearing in. If there is no iron pipe, the inner ring of the bearing can also be countered with iron bar, and the bearing can be horizontally sleeved into the shaft by striking symmetrically and gently.

(2) Hot charging. If the fit is tight, in order to avoid cracking or damaging the fit surface of the bearing inner ring, the hot mounting method can be used. First, put the bearing in the oil pan (or oil groove) for heating, keep the oil temperature at about 100℃, the bearing must be immersed in the oil, and can not contact with the bottom of the pan, the bearing can be lifted by wire. Heat evenly, take out the bearing after 30~40min, and push the bearing to the journal quickly while it is hot. In rural areas, the bearing can be heated on a 100W bulb, and then it can be sleeved on the shaft after 1h.

2.1.10　Task Summary

The insulation performance of the motor shell with good performance is good. When the insulation of the shell becomes poor, leakage will occur. At this time, it is necessary to cut off the power supply of the motor for detection and maintenance.

The detection method is to use a megger to detect whether the winding is insulated or not, and whether the shell is insulated or not. If the resistance between the windings or between the winding and the housing is not infinite, the motor is faulty.

The main causes of the leakage of the motor case are that the insulation of an outgoing line in the motor is damaged and touches the case; the local burning of the motor winding causes the leakage between the stator and the case. At this time, the internal circuit needs to be removed and repaired.

If it is in a high humidity environment, the motor will be damped and the insulation will be reduced, so that the shell will be charged. The stator winding of the motor shall be baked and dehumidified.

This task mainly studies the structure of the three-phase squirrel cage asynchronous motor, the working principle of the three-phase asynchronous motor, the nameplate data of the three-phase asynchronous motor, the inspection of the three-phase squirrel cage asynchronous motor, the disassembly and assembly of the three-phase asynchronous motor, which lays the foundation for the later study of the design and installation of the three-phase asynchronous motor circuit.

Task 2.2　Installation and Debugging of Starting Circuit of Three-phase Asynchronous Motor

2.2.1　Task Description

In industrial and agricultural production, electric motors are used to drive the operation of production equipment and convert electric energy into mechanical energy. So, how to control the starting and running of the motor in the production of industry and agriculture? Which auxiliary devices should be used? What is the structure and working principle of these auxiliary devices? How to use it?

2.2.2　Task Target

(1) Familiar with the structure and working principle of various low voltage electrical appliances.

(2) Master the use of various low-voltage electrical appliances.

(3) Master the wiring, checking and operation of the direct start control circuit of three-phase asynchronous motor.

(4) Master the control circuit principle of three-phase asynchronous motor inching, self-locking starting, inching and self-locking.

(5) Master the principles of short circuit protection, overload protection, voltage loss protection and undervoltage protection.

2.2.3 Task Analysis

The process of motor accelerating from static state to stable state after power on is called motor starting.

Direct starting is also called full voltage starting. It is a method of starting the motor by directly adding rated voltage to the stator winding of the motor through switches or contactors.

To control the starting of the motor, it is necessary to apply it to various low-voltage electrical appliances, mainly including circuit breakers, buttons, AC contactors, thermal relays, fuses, etc.

The starting control circuit of motor mainly includes manual control circuit, inching control circuit, self-locking control circuit, overload protection control starting circuit and control circuit which can operate continuously and inching control.

These control lines are various, but no matter simple or complex, they are generally composed of some basic control links. When analyzing the principle of control lines and judging their faults, they are generally started from these basic control links.

Therefore, it is of great help to analyze the working principle and maintenance of the whole electrical control circuit of the production machinery from learning various low-voltage electrical appliances to analyzing the basic control circuit composed of them.

This task mainly studies the manual control circuit, inching control circuit, self-locking control circuit, overload protection control starting circuit and control circuit that can operate continuously and inching control.

It is divided into five sub tasks:

(1) Know low voltage electrical appliances;
(2) Manual forward rotation control circuit of three-phase asynchronous motor;
(3) Three phase asynchronous motor inching control circuit;
(4) Three phase asynchronous motor self-locking control circuit;
(5) Three phase asynchronous motor overload protection control circuit.

2.2.4 Task-related Knowledge

2.2.4.1 Know Low-voltage Electrical Appliances

All electrical equipment that controls, regulates, detects, converts and protects the production, transmission, distribution and use of electric energy can be called electrical appliances. Low voltage electrical appliances are used for circuits under AC 1200V. and DC 1500V, which play the role of on-off, control, protection and regulation.

(1) Classification of low voltage apparatus. Low voltage electrical appliances have many functions, wide uses and various specifications. In order to master them systematically, they must be classified.

1) According to the action nature of electrical appliances.

① Manual electrical appliances: the electrical appliances that send out action instructions, such as knife switch, button, etc.

② Automatic electrical appliances: electrical appliances, such as contactors, relays, solenoid valves, etc., that can automatically complete the task of connecting and breaking circuits according to the electric or non electric signals without direct manual operation.

2) By purpose.

① Control electrical appliances: electrical appliances used for various control circuits and control systems, such as contactors, relays, motor starters, etc.

② Power distribution equipment: electrical equipment used for power transmission and distribution, such as knife switch, low-voltage circuit breaker, etc.

③ Master: used for sending action command in automatic control system, such as button, change-over switch, etc.

④ Protective electrical appliances: electrical appliances used to protect circuits and electrical equipment, such as fuses, thermal relays, etc.

⑤ Actuator: an electrical appliance used to perform a certain action or transmission function, such as an electromagnet, an electromagnetic clutch, etc.

3) According to working principle.

① Electromagnetic electrical appliances: electrical appliances that work according to the principle of electromagnetic induction. Such as AC and DC contactors, various electromagnetic relays, etc.

② Non electric control electric appliance: an electric appliance whose work depends on the change of external force or some non electric physical quantity. Such as knife switch, speed relay, pressure relay, temperature relay, etc.

(2) Circuit breaker. Circuit breaker refers to the switch device that can close, carry and open the current under the normal circuit condition and can close, carry and open the current under the abnormal circuit condition within the specified time. The circuit breaker is divided into high-voltage circuit breaker and low-voltage circuit breaker according to its scope of use. The demarcation of high-voltage and low-voltage boundary is relatively vague. Generally, the circuit breaker with more than 3kV is called high-voltage apparatus.

The circuit breaker can be used to distribute electric energy, start asynchronous motor infrequently, and protect power lines and motors. When they have serious overload, short circuit, undervoltage and other faults, it can cut off the circuit automatically. Its function is equivalent to the combination of fuse switch and over/under heat relay. It is not necessary to change the parts after breaking the fault current. At present, it has been widely used.

The circuit breaker is generally composed of contact system, arc extinguishing system, operating mechanism, release, shell, etc.

The miniature circuit breaker is shown in Figure 2-16. Miniature circuit breaker, commonly known as air switch, is a kind of plastic case circuit breaker. It is one of the most widely used terminal protection appliances in building electrical terminal distribution equipment. It is used for short-circuit, overload, over-voltage protection of single-phase and three-phase under 125A, in-

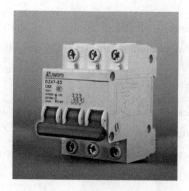

Figure 2-16 Miniature circuit breaker

cluding single pole 1P, two pole 2P, three pole 3P, four pole 4P.

(3) Fuse. Fuse refers to a kind of electrical appliance which breaks the circuit by fusing the melt with the heat generated by itself when the current exceeds the specified value. Fuse is a kind of current protector which is made according to the principle that after the current exceeds the specified value for a period of time, the fuse melts the melt with the heat generated by itself, thus breaking the circuit. Fuse is widely used in high and low voltage distribution system, control system and electrical equipment. As the protector of short circuit and over-current, fuse is one of the most commonly used protective devices.

1) Structure and function of fuse. The structure of fuse is generally divided into melt seat, melt and other parts. The fuse is connected in series in the protected circuit. When the circuit current exceeds a certain value, the melt will fuse due to heating, which will cut off the circuit and play a protective role. The heat of the melt is directly proportional to the square of the current passing through the melt and the continuous power on time. When the circuit is short circuited, the current is very large, the melt heats up rapidly, and it will fuse immediately. When the current value in the circuit is equal to the rated current of the melt, the melt will not fuse. So fuse can be used for short circuit protection. Because the heat can be accumulated by the overload current of the melt when the electric equipment is overloaded, and the heat accumulated by the melt after the electric equipment is overloaded for a certain period of time can also make it fuse, so the fuse can also be used as overload protection. See Figure 2-17 for the electrical symbols of fuse and Figure 2-18 for the outline of common fuse.

Figure 2-17 Fuse electrical symbols

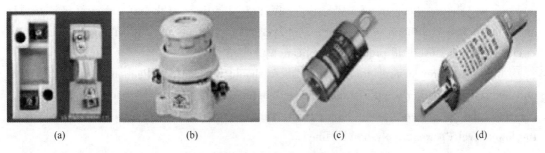

(a)　　　　　　　　(b)　　　　　　　　(c)　　　　　　　　(d)

Figure 2-18　Fuse outline

(a) Porcelain insert type; (b) Screw type; (c) Packing free tube type; (d) Packing sealed tube type

Among them, screw type fuse and ceramic plug-in fuse are commonly used in industrial production. They have the same function and are used for overload and short circuit protection of electrical equipment. High breaking capacity, compact structure, small volume, small installation area, convenient to change melt, safe and reliable operation.

The fuse tube is equipped with fuse and quartz sand, and there is a fuse indicator on it to indicate whether the fuse is broken. Quartz is used to enhance the arc extinguishing performance.

2) Selection of fuse. The requirements for fuse are: when the electrical equipment is in normal operation, the fuse shall not be fused; when there is a short circuit, it shall be fused immediately; when the current changes normally (such as the motor starting process), the fuse shall not be fused; when the electrical equipment is continuously overloaded, it shall be fused later. The selection of fuse mainly includes the selection of type and the determination of rated current of melt.

When choosing the type of fuse, it mainly depends on the protection characteristics of load and the size of short-circuit current. For example, fuses used to protect lighting and motors are generally considered for their overload protection. At this time, it is hoped that the melting coefficient of the fuses will be smaller.

Therefore, RC1A series fuses with lead-zinc alloy melt should be used for lighting lines and motors with small capacity, while for lighting lines and motors with large capacity, besides overload protection, the ability of breaking short-circuit current during short-circuit should also be considered. If the short circuit current is small, RC1A series fuses with tin melt or RM10 series fuses with zinc melt can be used. For the protection fuse of low-voltage power supply line in the workshop, the breaking capacity in case of short circuit is generally considered. When the short-circuit current is large, RL1 series fuse with high breaking capacity should be used. When the short-circuit current is quite large, RT0 series fuse with limited current action should be used.

The rated voltage of fuse shall be greater than or equal to the rated voltage of circuit.

The rated current of fuse shall be selected according to the load.

① For resistive load or lighting circuit, the starting process of such load is very short and the running current is relatively stable. Generally, the rated current of the melt is selected according to 1~1.1 times of the rated current of the load, and then the rated current of the fuse is selected.

② For inductive load such as motor, the starting current of such load is 4~7 times of the rated current, and the rated current of melt is generally 1.5~2.5 times of the rated current of motor. In general, the fuse is difficult to play the role of overload protection, but can only be used as short-circuit protection, overload protection application thermal relay.

③ In order to prevent over grade fusing, there should be good coordination between the upper and lower level (power supply trunk line and branch line) fuses. Therefore, the melt rated current of the upper level (power supply trunk line) fuse should be 1~2 levels higher than that of the lower level (power supply branch line).

(4) Contactor. Contactor is a kind of low-voltage control appliance which is widely used in electrical control system. It is used to connect and break AC and DC main circuits and large capacity control circuits frequently. The main control object is motor, which can realize remote control and has under (zero) voltage protection.

1) Structure and working principle. The contactor is mainly composed of electromagnetic system, contact system and arc extinguishing device. The outline and structure diagram are shown in Figure 2-19 and Figure 2-20.

Figure 2-19 Outline of AC contactor

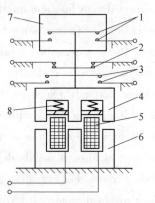
Figure 2-20 Structure diagram of contactor
1—Main contact; 2—Normally closed auxiliary contact;
3—Normally open auxiliary contact; 4—Moving core; 5—Electromagnetic coil;
6—Static core; 7—Arc extinguishing device; 8—Spring

Working principle: according to the electromagnetic working principle, when the electromagnetic coil is powered on, the coil current will generate a magnetic field, which will make the static core generate electromagnetic attraction to attract the armature, and drive the contact action to open the normally closed contact and close the normally open contact. The two are linked. When the electromagnetic coil is de energized, the electromagnetic force disappears, and the armature is released under the action of the release spring to restore the contact, that is, the normally open contact is open and the normally closed contact is closed. The graphic symbols and text symbols of the contactor are shown in Figure 2-21.

2) AC contactor. Contactor can be divided into AC contactor and DC contactor according to the type of main circuit current controlled by its main contact.

Among them, AC contactor coil is connected with AC, and main contact is connected and dis-

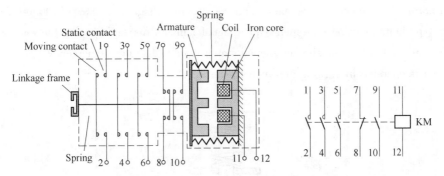

Figure 2-21 Structure and symbol of AC contactor

1~6—Three groups of normally open main contacts are set in terminals; 7, 8—Normally closed auxiliary contacts are set in terminals; 9, 10—Normally open auxiliary contacts are set in terminals; 11, 12—Control coil is set in terminals

connected with AC main circuit. DC contactor coil is connected with DC current, and main contact is connected and cut off DC main circuit.

As AC contactors are widely used, we only introduce AC contactors.

The structure and symbols of AC contactor are shown in Figure 2-21. In Figure 2-21, 1, 2, 3, 4, 5 and 6 are main contacts, 7 and 8 are normally closed auxiliary contacts, 9 and 10 are normally open auxiliary contacts, and 11 and 12 position coils. The right side of the drawing is its symbol in the drawing.

Working principle of AC contactor:

The AC contactor is mainly composed of main contact, auxiliary contact and control coil. When the control coil is powered on, the coil generates magnetic field, which attracts the armature through the iron core, while the armature drives all the moving contacts to act through the connecting rod, and contacts or disconnects with the static contacts. The current allowed to flow through the main contact of the AC contactor is larger than that of the auxiliary contact, so the main contact is usually connected to the main circuit with large current, and the auxiliary contact is connected to the control circuit with small current. The working principle of AC contactor is to make use of electromagnetic force and spring force to realize the connection and disconnection of contact. There are two working states of AC contactor: power loss state (release state) and power on state (action state). When the suction coil is powered on, the static iron core will generate electromagnetic attraction, the armature will be drawn in, and the connecting rod connected with the armature will drive the contact to act, so that the normally closed contact breaking contactor will be powered on; when the suction coil is powered off, the electromagnetic attraction will disappear, the armature will be reopened, the normally open contact will be closed, and the position spring will release, all contacts will be reset, and the contactor will be powered off.

(5) Control button. The control button is usually used as a switch to turn on or off the small current control circuit for a short time. The control button is composed of a button cap, a return spring, a bridge contact, a shell, etc. Generally, the composite structure with normally open con-

tact and normally closed contact is made, and the structure diagram is shown in Figure 2-22. The indicator type button can be equipped with a signal lamp to display signals; the emergency type button is equipped with a mushroom shaped button cap to facilitate emergency operation. Knob type button is operated by turning the knob by hand.

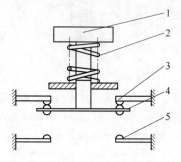

Figure 2-22 Structure of control button
1—Button cap; 2—Return spring; 3—Normally closed contact; 4—Moving contact; 5—Normally open contact

Button cap has many colors, generally red is used as stop button and green is used as start button. The buttons are mainly selected according to the number of contacts, use occasions and colors.

Its appearance is shownin Figure 2-23. Button is a kind of commonly used control electrical component, which is often used to connect or disconnect 'control circuit' (in which the current is very small), so as to achieve the purpose of controlling the operation of motor or other electrical equipment. The action law of the contact is: when it is pressed, the moving contact is opened first, then closed; when it is released, the moving contact is opened first, then closed.

Figure 2-23 Button

The graphic symbols and text symbols of the button switch are shown in Figure 2-24.

(6) Thermal relay. The thermal relay works on the principle of thermal effect of current, which is used for overload protection and phase loss protection of motor or other electrical equipment. However, due to the thermal inertia of the heating element in the thermal relay, the instantaneous overload protection and short-circuit protection cannot be done in the circuit, so it is different from the current relay and fuse.

1) Structure of thermal relay. The Structure of thermal relay is shown in Figure 2-25.

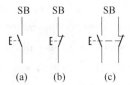

Figure 2-24 Graphic symbols and text symbols
(a) Normally open contact; (b) Normally closed contact; (c) Compound contact

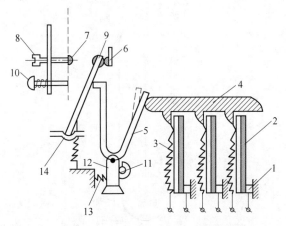

Figure 2-25 Structure of thermal relay

1—Terminal; 2—Bimetal plate; 3—Thermal element; 4—Guide plate; 5—Compensation bimetal plate;
6—Normally closed static contact; 7—Normally open contact; 8—Reset screw; 9—Normally closed moving contact;
10—Reset button; 11—Adjustment knob; 12—Support; 13—Compression spring; 14—Push rod

2) How it works. After the overload current passes through the thermal element, the bimetallic sheet is heated and bent to push the action mechanism to drive the contact action, so as to disconnect the motor control circuit to realize the motor power-off and stop, and play the role of overload protection. In view of the fact that the heat transfer takes a long time in the process of bimetal heating and bending, the thermal relay can only be used as overload protection instead of short circuit protection. The symbols of thermal relay are shown in Figure 2-26.

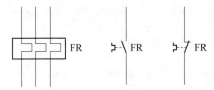

Figure 2-26 Symbols of thermal relay

2.2.4.2 Manual Forward Rotation Control Circuit of Three-Phase Asynchronous Motor

The manual forward rotation control circuit can only control the one-way start and stop of the motor. And drive the movement of production machinery to rotate or move in one direction.

For small capacity motors, as long as the motor is connected to the rated voltage, it can be started directly. This starting mode is called full pressure starting. For three-phase squirrel cage asynchronous motor, the current flowing through the motor during full voltage starting will far exceed the rated current of the motor, about 5~7 times of the rated current. Too large starting current will cause a large voltage drop on the line, which will affect the normal operation of other electrical equipment. Therefore, full voltage starting is only suitable for motors with small capacity. For motors with large capacity, the method of reduced pressure starting should be adopted.

The manual forward rotation control circuit is shown in Figure 2-27.

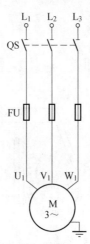

Figure 2-27 Manual forward rotation control circuit

(1) Characteristics of the circuit.

Advantages: simple and low cost.

Disadvantages: it can't be operated remotely and it's not safe to operate, so it's easy to burn hands.

(2) Functions of low-voltage apparatus. The manual switch QS is used to turn on and off the power supply; the fuse is used for short circuit protection.

(3) Working principle of the circuit.

Start: close the tool switch QS→Motor M is connected to the power supply to start operation.

Stop: disconnector QS→Motor M is disconnected from the power supply and stops running.

2.2.4.3 Inching Control Circuit of Three-phase Asynchronous Motor

Inching control circuit is the simplest control circuit to control motor with button and contactor. Schematic diagram is shown in Figure 2-28.

Inching control refers to: press the button, the motor is powered on for operation; release the button, the motor is powered off and stops running. This control method is commonly used in the hoist motor of the electric hoist and the motor control of the rapid movement of the lathe carriage.

The control circuit is usually represented by the electrical graphic symbols and text symbols specified in the national standards and drawn into the control circuit schematic diagram. It is based on the physical wiring circuit, used to express the working principle of the control circuit.

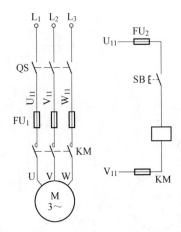

Figure 2-28 Schematic diagram of inching control circuit

The inching control schematic can be divided into two parts: the main circuit and the control circuit.

The main circuit is the circuit from the power supply L_1, L_2, L_3 to the motor M through the main contact of the power switch QS, fuse FU_1, contactor km. It flows through a large circuit. The control circuit is composed of fuse FU_2, button SB and the coil of contactor KM, and the current flowing is small. When the motor needs to be inched, first close the power switch QS, press the inching button sb, the contactor coil KM will be electrified, the armature will be closed, driving its three pairs of normally open main contacts km to close, and the motor M will be connected to the power supply for starting operation. After the SB button is released, the contactor coil is de energized, and the armature is reset under the action of spring force, driving its three pairs of normally open main contacts to open, and the motor is de energized and stopped.

Working principle of inching control circuit:

After closing the power switch QS.

Start: Press SB→KM coil energized→KM Main contact closed→Motor running

Stop: Release SB→KM coil power off→KM main contact open→Motor stop

2.2.4.4 Self Locking Control Circuit of Three-phase Asynchronous Motor

In order to realize the continuous operation of the motor, the forward rotation control circuit with contactor self-locking can be used. A normally open auxiliary contact of the contactor needs to be connected in parallel at both ends of the start button SB_2. In the control circuit, a stop button SB_1 can be connected in series to stop the motor. The motor self-locking control circuit is shown in Figure 2-29.

The forward control circuit of the self-locking contactor can not only make the motor run continuously, but also has the functions of under voltage protection and loss of voltage (zero voltage) protection.

(1) Undervoltage protection. 'Undervoltage' means that the line voltage is lower than the rated voltage to be applied to the motor. 'Undervoltage protection' refers to the protection that when the

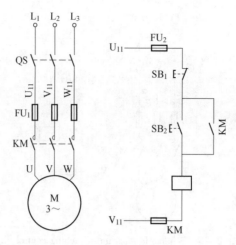

Figure 2-29 Motor self-locking control circuit

line voltage is lower than a certain value, the motor can automatically break away from the power supply voltage and stop running to avoid the motor running under undervoltage. Why do motors have undervoltage protection? When the power supply voltage drops, the current of the motor will rise. The more serious the voltage drop is, the more serious the current will rise. In the forward rotation control circuit of contactor self-locking, when the motor is running, the power supply voltage is reduced to a lower level (generally below 85% of the working voltage), the magnetic flux of contactor coil becomes weak, the electromagnetic attraction is insufficient, the moving core is released under the action of the reaction spring, the self-locking contact point is disconnected, and the self-locking is lost, at the same time, the main contact is also disconnected, and the motor stops running, which is protected.

(2) Loss of voltage (or zero voltage) protection. 'Zero voltage protection' means that when the motor is running, the power supply of the motor can be cut off automatically when the power supply is temporarily cut off due to some external reasons. When the power supply is restored, the motor can not be started by itself. If no preventive measures are taken, it is easy to cause personal accidents. The forward rotation control circuit with contactor self-locking is adopted. Since the self-locking contact and the main contact have been disconnected together when the power is cut off, neither the control circuit nor the main circuit will be connected by themselves. Therefore, when the power supply is restored, if the button is not pressed, the motor will not start by itself.

(3) Principle analysis of self-locking control circuit of three-phase asynchronous motor. Close QS.

Start:

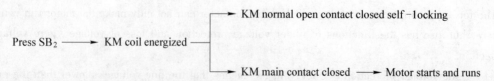

Stop:

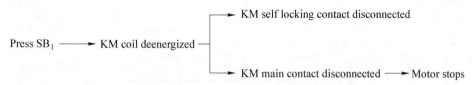

2.2.4.5 Self Locking Forward Rotation Control Circuit with Overload Protection

Overload protection is a kind of protection that can automatically cut off the power supply of the motor and make the motor stop running when the motor is overloaded. The most commonly used is to use thermal relay for overload protection. During the operation of the motor, such as long-term overload, frequent operation or open phase operation, the current of the motor stator winding may exceed its rated value, but the current does not reach the fuse, which will cause the overheating temperature of the motor stator winding to rise. If the temperature exceeds the allowable temperature rise, the insulation will be damaged, the service life of the motor will be greatly shortened, and even the motor will be burnt in serious cases.

Therefore, overload protection measures must be taken for the motor.

During the operation of the motor, the current exceeds the rated value due to overload or other reasons. After a certain period of time, the thermal elements of the thermal relay FR connected in series in the main circuit are bent by heating. Through the action mechanism, the FR normally closed contact connected in series in the control circuit is disconnected, the control circuit is cut off, the coil of the contactor KM is cut off, the main contact is disconnected, and the motor M stops running, reaching the over limit for the purpose of protection.

The self locking forward rotation control circuit with overload protection is shown in Figure 2-30.

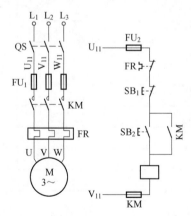

Figure 2-30 Self locking forward rotation control circuit with overload protection

2.2.5 Bill of Materials

The table of the bill of materials is shown in Table 2-5.

Table 2-5 Bill of materials

Serial number	Name	Number
1	Xk-SX2C advanced maintenance electrician training platform	1 station
2	Maintenance electrician training module (XKDT11)	1 pieces
3	Maintenance electrician training module (XKDT12A)	1 pieces
4	Three phase asynchronous motor	1 station
5	Jumper	Some
6	Multimeter	Some
7	Contactor	Some
8	Thermal relay	Some
9	Button	Some

2.2.6 Task Implementation

The table of the task implementation is shown in Table 2-6.

Table 2-6 Task implementation

Task specification of self-locking forward rotation control circuit with overload protection					
1. According to the requirements, complete the required device inventory and complete the following table					
Serial number	Device name	Number of devices	Device parameters	Device measurement	Remarks
2. Description of working principle					

Continued Table 2-6

Task specification of self-locking forward rotation control circuit with overload protection
3. Design and drawing of circuit wiring diagram
4. Operation process record
5. Power on test run process record
6. Task summary

2.2.7 Task Evaluation

The table of the task evaluation is shown in Table 2-7.

Table 2-7 Task evaluation

Scoring table for installation and commissioning of self-locking forward rotation control line with overload protection

Device inventory and measurement (10 points)

Serial Number	Key inspection contents	Scoring criteria	Score	Score	Remarks
1	Device inventory	Deduct 1 point for each error in device counting	5		
2	Device measurement	Deduct 1 point for each item of device measurement error	5		
		Subtotal			

Circuit diagram design (10 points)

Serial Number	Key inspection contents	Scoring criteria	Score	Score	Remarks
1	Main circuit design	Deduct 1 point for each error of main circuit	5		
2	Control circuit design	Deduct 1 point for each design error of control circuit	5		
		Subtotal			

Description of working principle (10 points)

Serial Number	Key inspection contents	Scoring criteria	Score	Score	Remarks
1	Working principle of main circuit	Deduct points if complete	5		
2	Operation principle of control circuit	Deduct points if complete	5		
		Subtotal			

Circuit installation (30 points)

Serial Number	Key inspection contents	Scoring criteria	Score	Score	Remarks
1	Whether the wire treatment is correct	Deduct 1 point for one mistake	10		
2	Whether the wire is firmly installed	Deduct 1 point for one mistake	10		
3	Is the circuit correct	Deduct 1 point for one mistake	10		
		Subtotal			

Power on test run (30 points)

Serial Number	Key inspection contents	Scoring criteria	Score	Score	Remarks
1	Self locking function	Whether the function is realized	10		
2	Overload protection function	Whether the function is realized	10		
3	Stop function	Whether the function is realized	10		
		Subtotal			

Continued Table 2-7

Scoring table for installation and commissioning of self-locking forward rotation control line with overload protection

Professionalism (10 points)

Serial Number	Key inspection contents	Scoring criteria	Score	Score	Remarks
1	Live operation	Deduct 5 points at a time until all points are deducted			
2	Standard operation	Deduct 1 point at a time until all points are deducted			
3	team work	As appropriate			
4	Clean working position	As appropriate			
	Total				

2.2.8 Troubleshooting

When troubleshooting the motor control circuit, make reasonable use of tools, such as multimeter. You can use the AC voltage range of the multimeter AC 500V to measure whether the power supply is normal, or use the ohm range of the multimeter to measure whether the circuit and low-voltage appliance are connected (Note: power should be cut off at this time).

The main faults in the installation and commissioning of the starting control circuit of the motor are as follows:

Fault 1, In the inching control circuit, close the switch QS, press the button SB, the contactor KM has no action, and the motor does not run.

Troubleshooting: disconnect the power supply, use the universal ohm gear, and measure whether the five fuses in the circuit are normal. If they are connected, measure whether the wires and QS are connected one by one. If they are still normal, turn the multimeter to the AC voltage range AC 500V, and measure whether the power supply voltage is normal.

Fault 2, In the self-locking control circuit, close the switch QS, press the button SB_2 once, the contactor KM is closed, the motor runs, release the SB_2, the KM loses power and opens, the motor stops running.

Troubleshooting: the fault is that the contactor km is not self-locking, resulting in the motor can not operate continuously. At this time, use the ohm gear of the multimeter to check whether the up in and down out wiring of the normally open contact of the contactor in parallel with SB_2 is normal.

Fault 3, In the inching control circuit, press button SB once, the contactor is closed, the motor is powered on to start operation, release the button, the main contact of contactor KM is still on, and the motor is still running.

Troubleshooting: the fault is that the main contact of the contactor is aged and rusty, and the action is not effective. After the contactor coil loses power, the main contact cannot be separated, resulting in the motor still running with power on. The maintenance method is to disassemble the contactor and grind the main contact and auxiliary contact of the contactor with file.

Pay attention to the following items during the line installation to reduce the occurrence of faults:

(1) When wiring, pay attention to centralized wiring, reduce overhead and crossing, and make horizontal, vertical and turning into a right angle.

(2) During operation, it is not allowed to touch the conductive part of each electrical component and the rotating part of the motor by hand to avoid electric shock and accidental damage.

(3) Only when the power is cut off, can theohm gear of the multimeter be used to check whether the wiring is correct.

(4) The wiring of screw type fuse shall be correct to ensure the safety of power consumption.

(5) The training shall be completed within the specified time, and safe operation and civilized production shall be achieved at the same time.

The table of the troubleshooting record is shown in Table 2-8.

Table 2-8 Troubleshooting Record

	Troubleshooting record	
Fault 1	Fault Phenomenon	
	Cause of fault	
	Troubleshooting Process	
Fault 2	Fault Phenomenon	
	Cause of fault	
	Troubleshooting Process	
Fault 3	Fault Phenomenon	
	Cause of fault	
	Troubleshooting Process	
Fault 4	Fault Phenomenon	
	Cause of fault	
	Troubleshooting Process	

2.2.9 Task Development

Continuous operation and inching control mixed control circuit:

When the machine tool equipment is in normal operation, it is generally required that the motor is in continuous operation state, but when testing or adjusting the relative position of the cutter and the workpiece, it is required that the motor can be inching controlled. The line to realize this process requirement is the forward control line of continuous and inching hybrid control. Their main circuits are the same.

Figure 2-31 shows the continuous and inching control circuit.

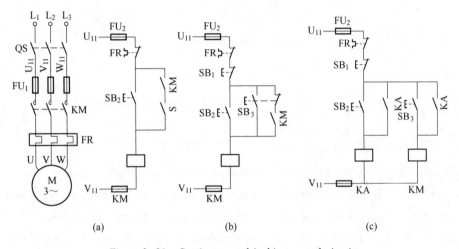

Figure 2-31　Continuous and inching control circuit

Figure 2-31 (a) is based on the forward control circuit of the contactor, and a manual switch S is connected in series in the self-locking circuit.

Working principle: first turn on the power switch QS, when S is off, press SB_2, which is inching control; when S is on, press SB_2, which is continuous control with self-locking.

Figure 2-31 (b) is a compoundbutton SB_3 added to the control circuit with self-locking.

Wheninching operation is required, press the SB_3 inching button, and its normally closed contact will first open the self-locking circuit, then close and connect the starting control circuit after the normally open trigger, the KM contactor coil will be powered on, the main contact will be closed, the three-phase power supply will be connected, and the motor will start and operate. When the inching button SB_3 is released, the KM coil loses power, the main contact of KM is disconnected, and the motor stops running.

If the motor needs to operate continuously, it is controlled by stop button SB_1 and start button SB_2, and the auxiliary contact of contactor KM acts as self-locking.

Figure 2-31 (c) shows that a jog button SB_3 and an intermediate relay KA are added to the control circuit.

Turn on the power switch QS first, Continuous control:

Start:

Click SB_2 ⟶ KA closed due to coil energization ⟶
- KA normally open contact in parallel with SB_2 closed and self locked
- KA normally open contact closed in parallel with SB_3 ⟶ KM is closed because the coil is energized ⟶ Motor M starts and runs

Stop:

Press SB_1 ⟶
- KA released due to coil power failure ⟶ KA two normal open contacts are open
- KM released due to coil power failure ⟶ Motor M power off and stop

Inching control:

Start: Press SB_3 ⟶ KM is closed because the coil is energized ⟶ Motor M starts and runs

Stop: Release SB_3 ⟶ KM released due to coil power failure ⟶ Motor M is powered off

The above three control circuits have their own advantages and disadvantages.

Figure 2-31 (a) is relatively simple. Since both continuous and inching are controlled by the same button SB_2, if the operation of switch S is neglected, it will cause confusion.

Although the continuous and jog buttons are separated in Figure 2-31 (b), when the contact center is released slowly due to residual magnetism, the jog control will become continuous control. For example, when the SB_3 is released, its normally closed contact should be closed after the KM self-locking contact is opened. If the contactor is released slowly, the KM self-locking contact has not been opened, the normally closed contact of SB_3 has been closed, and the KM coil will no longer be powered off and become continuous control. This is very dangerous in some limit states. So this kind of control circuit is simple but not reliable.

Figure 2-31 (c) diagram uses an intermediate relay KA, which is not economical, but its reliability is greatly improved.

2.2.10 Task Summary

This task learned the starting control circuit of low-voltage electrical appliances and motors.

There are many brands of low-voltage electrical appliances, such as circuit breakers, control buttons, contactors, thermal relays, fuses, etc.

The process of motor accelerating from static state to stable state after power on is called motor starting.

Direct starting is also called full voltage starting. It is a method of starting the motor by directly adding rated voltage to the stator winding of the motor through switches or contactors.

The starting control circuit of the motor mainly includes manual forward rotation control circuit, inching control circuit, self-locking control circuit, overload protection control starting circuit and control circuit that can operate continuously and inching control.

Task 2.3 Installation and Commissioning of Two-place Control Circuit of Three-phase Asynchronous Motor

2.3.1 Task Description

In order to ensure the safety of operation, some large-scale production equipment often requires multiple locations to perform control operations, such as the blower of the boiler room, the circulating water pump motor, and the machine tool, which need to be controlled locally on site and remotely in the control room. This task realizes the control of the motor through the contactor and the button.

2.3.2 Task Target

(1) Recognize the connection method and function of the main contact and auxiliary contact of the AC contactor.

(2) Understand the meaning, function and realization method of control in the two places.

(3) Learn the connection method and precautions of the startbutton and stop button in the control of the two places.

2.3.3 Task Analysis

Take the control of two places as an example. If a motor needs to realize the control of A and B, SB_{11} and SB_{12} are the start button and stop button installed in A place, SB_{21} and SB_{22} are the start button and stop button installed in B place. The characteristic of the circuit is that the start buttons SB_{11} and SB_{21} of the two places are connected in parallel; the stop buttons SB_{12} and SB_{22} are connected in series. In this way, the same motor can be started and stopped in both A and B to achieve the purpose of convenient operation.

2.3.4 Task-related Knowledge

The working principle of the control circuit of the three-phase asynchronous motor A and B: Press the start button SB_{11} or the start button SB_{21} of the ground, the contactor KM coil is energized, the main contact of the contactor KM is closed, and the contactor KM assists the normally open The point is closed, it acts as a self-locking, and the motor keeps running; press the stop button SB_{12} of ground A or stop button SB_{22} of ground B, the contactor KM coil is de-energized, the contactor KM main contact is opened, and the contactor KM auxiliary normally open contact Disconnect, cancel the self-locking effect, and the motor stops running.

Figure 2-32 is a circuit diagram of the control circuit of the three-phase asynchronous motor A and B. In the circuit, the start buttons SB_{11} and SB_{21} of A and B are connected in parallel, and the stop buttons SB_{12} and SB_{22} are connected in series. Therefore, both A and B can start and stop the same three-phase asynchronous motor.

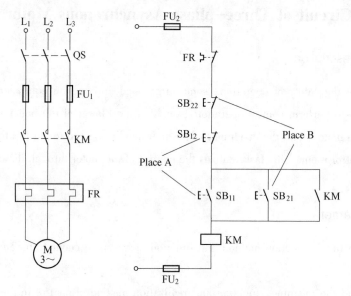

Figure 2-32 Three-phase asynchronous motor Place A and B control circuit diagram

QS—Circuit Breaker; FR—Thermal relay; FU—Fuse; KM—Contactor;

SB_{11}—Place A start button; SB_{12}—Place A stop button; SB_{21}—Place B start button; SB_{22}—Place B stop button

2.3.5 Bill of Materials

The table of the bill of materials is shown in Table 2-9.

Table 2-9 Bill of materials

No.	Name	Number
1	XK-SX2C Advanced Maintenance Electrician Training Platform	1
2	Repair Electrician Training Components (XKDT11)	1
3	Repair Electrician Training Components (XKDT12A)	1
4	Three-phase Asynchronous Motor	1
5	Jumper	Several

2.3.6 Task Implementation

The table of the task implementation is shown in Table 2-10.

Table 2-10　Task implementation

Task list for installation and commissioning of two-phase control circuit of three-phase asynchronous motor						
1. Complete the required device inventory and complete the following table as required							
No.	Device	Number	Parameters	measurement	Remarks

2. Working principle description

3. Design and drawing of circuit wiring diagram

Continued Table 2-10

Task list for installation and commissioning of two-phase control circuit of three-phase asynchronous motor
4. Operation process record
5. Recording process of power-on test drive
6. Task summary

2.3.7 Task Evaluation

The table of the task evaluation is shown in Table 2-11.

Table 2-11 Task evaluation

Installation of two-phase control circuit of two-phase asynchronous motor evaluation form					
Device inventory and measurement (10 points)					
No.	Key inspection	Grading	Total score	Score	Remarks
1	Device inventory	1 point for one error	5		
2	Device measurement	1 point for one error	5		
		Subtotal			
Circuit diagram design (10 points)					
No.	Key inspection	Grading	Total score	Score	Remarks
1	Main circuit design	1 point for one error	5		
2	Control circuit design	1 point for one error	5		
		Subtotal			

Continued Table 2-11

Installation of two-phase control circuit of two-phase asynchronous motor evaluation form

Working principle description (10 points)

No.	Key inspection	Grading	Total score	Score	Remarks
1	Working principle of main circuit	Complete discretion	5		
2	Working principle of control circuit	Complete discretion	5		
		Subtotal			

Line construction (30 points)

No.	Key inspection	Grading	Total score	Score	Remarks
1	Whether the wire is handled correctly	1 point for one error	10		
2	Whether the wire is firmly installed	1 point for one error	10		
3	whether the line correct	1 point for one error	10		
		Subtotal			

Power on test (30 points)

No.	Key inspection	Grading	Total score	Score	Remarks
1	Features of Place A	Whether the function is realized	15		
2	Features of Place B	Whether the function is realized	15		
		Subtotal			

Professional accomplishment (10 points)

No.	Key inspection	Grading	Total score	Score	Remarks
1	Power on operation	5 point for one error			
2	Standard operation	1 point for one error			
3	Team work	Discretion			
4	Work place tidy	Discretion			
		Total			

2.3.8 Troubleshooting

2.3.8.1 Common Faults

(1) Press SB_{11}, SB_{21} motor can not start.

(2) The motor can only be jogged.

(3) Pressing SB_{11} motor cannot start, pressing SB_{21} motor can start.

2.3.8.2 Troubleshooting Flowchart

The troubleshooting flowchart is shown in Figure 2-33.

The table of the troubleshooting record is shown in Table 2-12.

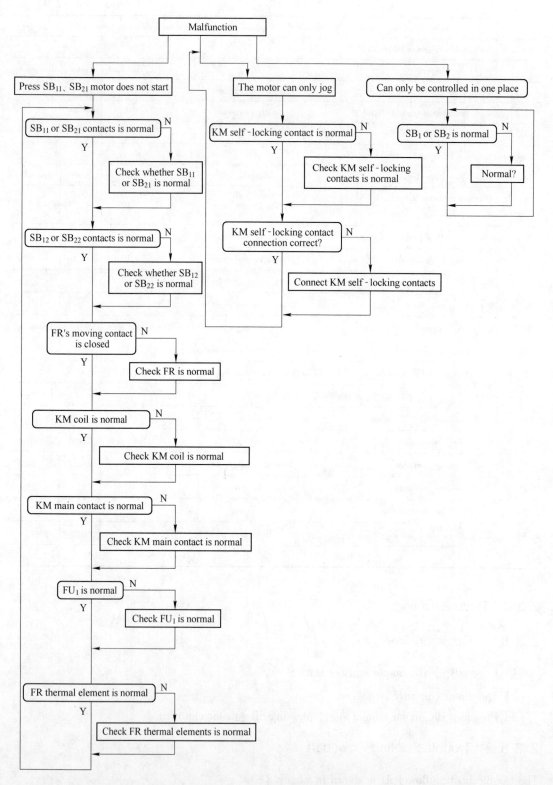

Figure 2-33 Troubleshooting flowchart

Table 2-12 Troubleshooting record

	Troubleshooting record	
Fault 1	Fault phenomenon	
	Cause of fault	
	Trouble shooting process	
Fault 2	Fault phenomenon	
	Cause of fault	
	Trouble shooting process	
Fault 3	Fault phenomenon	
	Cause of fault	
	Trouble shooting process	
Fault 4	Fault phenomenon	
	Cause of fault	
	Trouble shooting process	

2.3.9 Task Development

Three-phase asynchronous motor three-ground control circuit is realized. On the basis of the original two-ground control circuit, add a pair of buttons to design a 'three-phase asynchronous motor three-ground control circuit'.

Mission accomplished:

(1) Complete the design of the circuit diagram.

(2) Complete the principle explanation of 'three-phase asynchronous motor three-ground control circuit'.

2.3.10 Task Summary

The motor needs to be controlled in two places, and the start buttons SB_{11} and SB_{21} are connected in parallel, and the stop buttons SB_{12} and SB_{22} are connected in series. Therefore, both places can start and stop the same three-phase asynchronous motor.

During the implementation of the task, the power supply should be disconnected first, and the experimental wiring should be carried out according to the figure. First, power on the control loop to see whether the contactor is operating correctly. After the control loop is debugged correctly, then power on the main loop to see if the motor is running correctly. During the implementation process, it is not possible to carry out the wiring operation with power on. The control loop should be debugged first and then the main loop.

Project 3　Complex Control Skill Training of Three-phase Asynchronous Motors

Task 3.1　Installation and Debugging of Three-phase Asynchronous Motor Forward and Reverse Control Circuit

3.1.1　Task Description

Production machinery often requires that moving parts can move in both positive and negative directions, which requires that the driving motor can rotate in both positive and negative directions, such as the retractable door, elevator, etc. According to the principle of motor, changing the phase sequence of three-phase power supply of motor can change the direction of motor. In this task, the contactor and button are used to realize the forward and reverse control of electric motor.

3.1.2　Task Target

(1) Recognize the connection method between AC contactor and auxiliary contact and its function.

(2) Understand the meaning, function and method of interlock.

(3) Learn to realize various methods and precautions of motor forward and reverse.

3.1.3　Task Analysis

In order to realize the positive and reverse control of the motor, any two phases in the phase sequence of its power supply can be adjusted (we call it commutation), usually V phase is unchanged, and U phase and W phase can be adjusted relatively. In order to ensure that the phase sequence of the motor can be exchanged reliably when the two contactors act, the upper wiring of the contactor shall be consistent during wiring, and the lower phase of the contactor shall be adjusted. Since the two phases are in phase sequence, it is necessary to ensure that the two KM coils cannot be powered on at the same time, or serious phase to phase short circuit fault will occur, so interlocking must be adopted. For the sake of safety, the double interlock forward and reverse control circuit of button interlock (Mechanical) and contactor interlock (Electrical) is often used; when button interlock is used, even if the forward and reverse buttons are pressed at the same time, the two contactors for phase adjustment cannot be powered at the same time, and the mechanical short circuit between phases is avoided. In addition, due to the application of contactor interlocking, as long as one of the contactors is powered on, its long closed contact will not be closed, so in the application of mechanical and electrical double interlocking, the power supply system of the motor can not be short circuited between phases, effectively protecting the motor, at the same time, avoiding the accident caused by the short circuit between phases during phase adjustment, burning the contactor.

The task is dividedinto four sub tasks:

(1) Installation and debugging of forward and reverse control circuit of contactor interlocking;

(2) Installation and debugging of forward and reverse control circuit of button interlocking control motor;

(3) Installation and debugging of forward and reverse control circuit of contactor button double interlocking;

(4) Installation and debugging of forward and reverse control circuit of contactor button double interlocking.

3.1.4 Task-related Knowledge

(1) Working principle of three-phase motor: when the three-phase stator winding of the motor After the three-phase symmetrical alternating current is connected, a rotating magnetic field will be generated, which will cut the rotor winding and generate the induced current in the rotor winding (the rotor winding is a closed path). The current carrying rotor conductor will generate the electromagnetic force under the action of the stator rotating magnetic field, thus forming the electromagnetic torque on the motor shaft, driving the motor to rotate, and the motor rotation direction and rotation Phase in the direction of rotating magnetic field. If any two of the phase sequence of the power supply are adjusted relative to each other, the generated rotating magnetic field will change the direction, so that the rotation direction of the motor will also change, so as to realize the positive and negative rotation of the motor.

(2) Interlock: the 'forward' and 'reverse' buttons of the same motor in the forward and reverse control of three-phase motor shall realize interlock control, that is, when one button is pressed, the other button must automatically disconnect the circuit, so as to effectively prevent the mechanical failure or personal injury caused by the simultaneous energization of the two buttons. The application of interlock in motor is contactor interlock forward and reverse circuit, double interlock forward and reverse circuit, button interlock forward and reverse circuit.

3.1.4.1 Installation and Debugging of Forward and Reverse Control Circuit of Contactor Interlock

Figure 3-1 shows the forward and reverse control circuit of contactor interlock. Using forward contactor KM_1 and reverse contactor KM_2 to complete the two-phase power supply of the main circuit, so as to realize the conversion of forward and reverse.

In the control circuit, the normally closed contact KM_1 (④-⑤) of forward contactor KM_1 is used to control the coil of reverse contactor KM_2, and the normally closed contact KM_2 (②-③) of reverse contactor KM_2 is used to control the coil of forward contactor KM_1, so as to achieve mutual locking. These two pairs of normally closed contacts are called interlocking contacts, and the

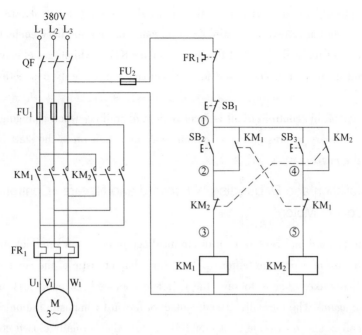

Figure 3-1 forward and reverse control circuit diagram of contactor interlock

QF—Circuit breaker; FR—Thermal relay; FU—Fuse; SB_1—Stop button; SB_2—Forward button; SB_3—Reverse button; KM_1—Forward contactor; KM_2—Reverse contactor

circuit composed of these two normally closed contacts is called interlocking link.

When the power switch is closed, press the forward start button SB_2, the forward contactor KM_1 coil is powered on and closed, the normally open main contact of the main circuit is closed, and the motor is started and operated in the forward direction. At the same time, the normally open auxiliary contact KM_1 (①-②) of the control loop is closed to realize self-locking; the normally closed auxiliary contact KM_1 (④-⑤) is open to cut off the coil circuit of the reverse contactor KM_2 to realize interlocking.

When it is necessary to stop, press the stopbutton SB_1, cut off the coil power of forward contactor KM_1, release the armature of contactor KM_1, normally open main contact recovers the off state, motor stops running, meanwhile, self-locking contact also recovers the off state, self-locking function is released, so as to prepare for the next start.

The reverse start process can be completed only by pressing the reverse start button SB_3, and the steps are similar to the forward start.

Function of interlock contact: suppose that after pressing forward start button SB_2 and forward start of motor, if reverse start button SB_3 is pressed again for some reason (such as misoperation), the reverse contactor will not be connected because interlock contact KM_1 (④-⑤) of forward contactor has been disconnected. Obviously, if there is no interlock function of the auxiliary contact KM_1 (④-⑤), the coil of the reverse contactor KM_2 will be energized, which will inevitably cause the six normally open contacts of the positive and reverse contactors of the main return circuit to be completely closed, resulting in the power short circuit accident, which is absolutely not allowed! Similarly, after reverse starting, the normally closed auxiliary contact of reverse contactor KM_2 will cut off the coil circuit of forward contactor KM_1, which can effectively prevent the power short circuit accident caused by the forward contactor mistakenly connecting the main circuit.

The disadvantage of this control circuit is inconvenient operation: when changing the direction of the motor, it is necessary to press the stop button first, and then press the start button to change the direction of the motor.

3.1.4.2 Installation and Debugging of Forward and Reverse Control Circuit of Button Interlock Control Motor

Figure 3-2 is the forward and reverse control circuit diagram of button interlocking control motor, which requires the use of composite buttons. The action characteristic of the compound button is to break first and then make, that is to say, the dynamic contact is to open first and the dynamic contact is to close again. The normally closed contact of forward composite button SB_2 is connected in series in the coil circuit of reverse contactor KM_2, while the normally closed contact of reverse composite button SB_3 is connected in series in the coil circuit of forward contactor KM_1. In this way, when SB_2 is pressed, only the coil of forward contactor KM_1 can be powered on and closed, while when SB_3 is pressed, only the reverse contactor KM_2 can be powered on and closed.

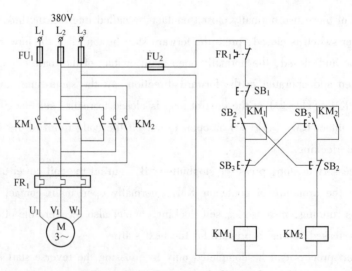

Figure 3-2 Button interlock control

In case of misoperation, for example, if two start buttons SB_2 and SB_3 are pressed at the same time, neither of the two contactors will be powered on and closed, which can prevent the short circuit accident of the main circuit caused by the simultaneous closing of the two contactors.

The advantage of this control circuit is convenient operation. When it is necessary to change the direction of the motor, it is not necessary to press the stop button first. However, this kind of circuit is also prone to short-circuit fault. For example, when the main contact of the contactor KM_1 is delayed or cannot be released due to some reasons, if the reversing button SB_3 is pressed at this time, the coil of the contactor KM_2 will be energized, and its main contact will be closed, and the positive and negative transfer contacts will be closed at the same time, resulting in a short circuit of two-phase power supply. Obviously, this kind of line is not safe enough.

3.1.4.3 Installation and Debugging of Contactor Button Double Interlock Forward and Reverse Control Circuit

When Figure 3-1 and Figure 3-2 are combined, the forward and reverse control circuit with contactor button double interlock is formed, as shown in Figure 3-3. The circuit combines the advantages of the two circuits mentioned above. It can start reversely without pressing the stop button and directly press the reverse button. When the forward transfer contact has fusion welding fault, it will not have phase to phase short circuit fault.

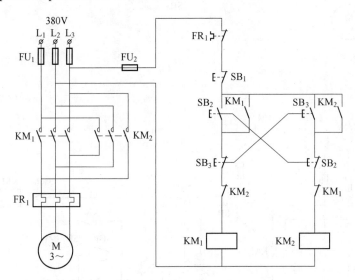

Figure 3-3 Forward and reverse motor control circuit diagram of button contactor double interlock

3.1.4.4 Three-phase Asynchronous Motor Forward and Reverse Inching and Starting Control Circuit

The circuit shown in Figure 3-4 (a) and Figure 3-4 (b) has reversible inching and reverse running, and is equipped with double interlock mechanism of button and contactor contact, which is convenient for operation. In Figure 3-4 (a) and Figure 3-4 (b), SB_1 and SB_2 are motor forward and reverse start buttons, SB_3 and SB_4 are motor forward and reverse inching buttons, SB_5 is stop button, and KM_1 and KM_2 are control motor forward and reverse AC contactors. In Figure 3-4 (b), there are only 4 control lines (①-④) between the button and the contactor and fuse, and 7 lines shown in Figure 3-4 (a). If the distance between the button and the controlled apparatus is very long, the wires shown in Figure 3-4 (b) can be used to save more wires.

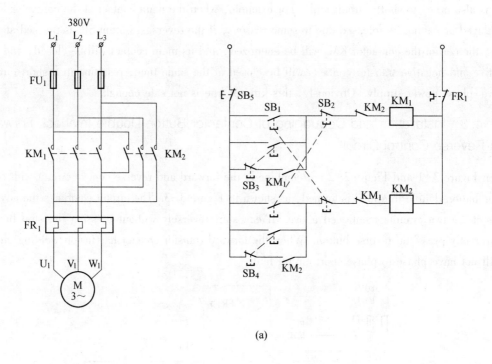

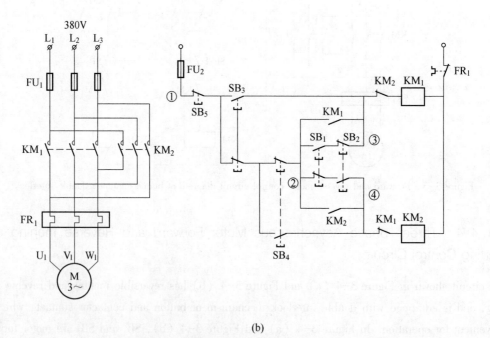

Figure 3-4 Forward and reverse inching and starting control circuit

3.1.5 Bill of Materials

The table of the bill of materials is shown in Table 3-1.

Table 3-1 Bill of materials

No.	Name	Number
1	XK-SX2C Advanced Maintenance Electrician Training Platform	1
2	Repair Electrician Training Components (XKDT11)	1
3	Maintenance electrician training component (XKDT12A)	1
4	Three phase asynchronous motor	1
5	Jumper	Several

3.1.6　Task Implementation

The table of the task implementation is shown in Table 3-2.

Table 3-2 Task implementation

( ) mission statement				
1. According to the requirements, complete the required devices and complete the following table				
Serial number	Device name	Number of devices	Device parameters	Device measurement	Remarks

2. Description of working principle

3. Design and drawing of circuit wiring diagram

Continued Table 3-2

( ) mission statement
4. Operation process record
5. Power on test run process record
6. Task summary

3.1.7 Task Evaluation

The table of the task evaluation is shown in Table 3-3.

Table 3-3 Task evaluation

( ) scoring table						
Device inventory and measurement (10 points)						
No.	Inspection contents	Scoring criteria	Partition	Score	Remarks
1	Device inventory	Deduct 1 point for each error	5		
2	Device measurement	Deduct 1 point for each error	5		
		Subtotal			
Circuit diagram design (10 points)						
No.	Inspection content	Scoring criteria	Partition	Score	Remarks
1	Main circuit design	Deduct 1 point for one mistake	5		
2	Control circuit design	Deduct 1 point for one mistake	5		
		Subtotal			
Description of working principle (10 points)						
No.	Inspection content	Scoring criteria	Partition	Score	Remarks
1	Working principle of main circuit	As appropriate	5		
2	Operation principle of control circuit	As appropriate	5		
		Subtotal			

Continued Table 3-3

( ) scoring table

Line construction (30 points)

No.	Inspection content	Scoring criteria	Partition	Score	Remarks
1	Whether the wire treatment is correct	Deduct 1 point for one mistake	10		
2	Whether the wire is firmly installed	Deduct 1 point for one mistake	10		
3	Is the line correct	Deduct 1 point for one mistake	10		
		Subtotal			

Power on test run (30 points)

No.	Inspection content	Scoring criteria	Partition	Score	Remarks
1	Forward rotation function	Whether the function is realized	10		
2	Reverse function	Whether the function is realized	10		
3	Interlock function	Whether the function is realized	10		
		Subtotal			

Professional quality (10 points)

No.	Inspection content	Scoring criteria	Partition	Score	Remarks
1	Live operation	Deduct 5 points at a time until all points are deducted			
2	Standard operation	Deduct 5 points at a time until all points are deducted			
3	team work				
4	Clean working position				
		Total			

3.1.8 Troubleshooting

General exclusion steps:

(1) Connect the three-phase power supply to the circuit control board and connect the motor;

(2) Check whether the wiring is correct, close the switch to conduct power on detection for the connected physical board; operate the forward start stop and reverse start, and observe the operation of the motor;

(3) In case of any fault, it is necessary to cut off the power supply for detection, analyze the cause of the fault and let the students retest, debug and power on again until the test run is successful;

(4) Finally, the failure phenomenon and its causes are evaluated.

Problems and solutions:

(1) No self-locking of control circuit. This is because the normally open of the AC contactor KM_1 (or KM_2) is not in parallel with the switch SB_2 (SB_3) and is connected in series with the

coil. When this problem occurs, check whether KM_1 has no self-locking or KM_2 has no self-locking. If KM_1, check whether KM_1 is normally open, otherwise check KM_2.

(2) Nointerlock of control circuit. This is because the normally closed of the two AC contactors KM_1 and KM_2 do not control each other's coil circuit, that is, the normally closed of KM_1 (KM_2) is not connected in series with the coil circuit of KM_2 (KM_1).

(3) Control circuit is not live. It may be caused by the control circuit not taking phase (phase return). At this time, check the control circuit. When the switch SB_2 or SB_3 is pressed, check whether there is a path. If there is a path, check whether the fuse is normal.

(4) Main circuit is not live. At this time, the switch may not be closed, or the fuse may be burnt out, or the main contact may be in poor contact, which can be measured with a multimeter, and then determine the problem.

(5) Lack of phase in circuit. It shows that the motor speed is slow and produces a large noise. At this time, it can measure the three-phase circuit, determine the less phase circuit and adjust it.

(6) Short circuit. This problem is the most serious, the whole circuit must be measured and checked.

The examples of common faults is shown in Table 3-4.

Table 3-4 Examples of common faults

Fault phenomenon	Cause of failure
The positive and negative rotation of the motor are in phase loss, and the winding of KM_1 and KM_2 are normal	The common circuit in KM_1 and KM_2 main circuit is damaged by circuit, contact, fuse or motor winding
Motor positive rotation phase loss, reverse normal KM_1, KM_2 coil pull in are normal	The circuit or contact of U-phase in KM_1 main circuit is damaged
The positive rotation of the motor is normal, the reverse rotation is lack of phase, and the coils of KM_1 and KM_2 are all closed normally	The circuit or contact of V-phase in the main circuit of KM_2 is damaged
There is no positive and negative rotation of the motor, and KM_1 and KM_2 coils are not closed	KM_1 and KM_2 coils shall not be electrified, and the common circuit in the control circuit has line, contact or fuse damage
The positive and reverse rotation of the motor are normal, KM_1 and KM_2 coils are closed, but the motor cannot be stopped	Short circuitin the upper and lower lines of SB_3 contact or SB_3 contact in the control circuit
The motor has no forward rotation, the reverse rotation is normal, KM_1 coil is not closed, and KM_2 coil is closed normally	The circuit, contact or coil in the forward rotation control circuit of the motor is damaged
Motor normal forward rotation, reverse inching, KM_1 coil pull in normal, KM_2 coil pull in inching	KM_2 control circuit is unable to self lock, there is line or contact damage in the self-locking circuit

Continued Table 3-4

Fault phenomenon	Cause of failure
The motor has normal forward rotation without reverse rotation, KM_1 coil is normally closed, and KM_2 coil is not closed	The circuit, contact or coil in the motor reverse control circuit is damaged

The table of the troubleshooting record is shown in Table 3-5.

Table 3-5 Troubleshooting record

	Troubleshooting record	
Fault 1	Fault phenomenon	
	Cause of fault	
	Troubleshooting process	
Fault 2	Fault phenomenon	
	Cause of fault	
	Troubleshooting process	
Fault 3	Fault phenomenon	
	Cause of fault	
	Troubleshooting process	
Fault 4	Fault phenomenon	
	Cause of fault	
	Troubleshooting process	

3.1.9 Task Development

Automatic round trip control line of worktable.

The travel switch is the main type of position switch. Its function is the same as the button. It can convert the mechanical signal into the electrical signal. Only the action of the contact does not depend on the manual operation, but uses the collision of the mechanical moving parts to make the contact action to realize the connection and disconnection of the circuit and achieve a certain control purpose. It is usually used to limit the position and stroke of mechanical movement, so that the moving machinery can stop automatically according to a certain position or stroke, reverse movement, variable speed movement or automatic back and forth movement, etc.

On the basis of the original forward and reverse circuit, adding stroke switches SQ_1 and SQ_2 to design a 'table automatic round trip control circuit'.

Design requirements:

(1) Realize the start stop and forward and reverse rotation functions of the motor;

(2) The stroke switch SQ_1 is triggered during the forward rotation of the motor, and the motor reverses;

(3) Trigger stroke switch SQ_2 motor forward rotation during motor reverse.

Complete the task:

(1) Complete the design of circuit diagram;

(2) Complete the explanation of the principle of 'table automatic round-trip control line'.

3.1.10 Task Summary

In order to realize the positive and reverse control of the motor, any two phases in the phase sequence of its power supply can be adjusted (we call it commutation), usually V phase is unchanged, and U phase and W phase can be adjusted relatively. In order to ensure that the phase sequence of the motor can be exchanged reliably when the two contactors act, the upper wiring of the contactor shall be consistent during wiring, and the lower phase of the contactor shall be adjusted. Since the two phases are in phase sequence, it is necessary to ensure that the two KM coils cannot be powered on at the same time, or serious phase to phase short circuit fault will occur, so interlocking must be adopted. The interlocking mode includes contactor interlocking, button interlocking, button contactor double interlocking, etc.

During the implementation of the task, the power supply shall be disconnected first, and the experimental wiring shall be carried out according to the drawing; the control circuit shall be electrified and debugged first to see whether the action of the contactor is correct; after the control circuit is debugged correctly, the main circuit shall be electrified and debugged to see whether the operation of the motor is correct. In the process of implementation, the wiring can not be carried out live. The main loop should be debugged after the control loop is debugged.

Task 3.2 Installation and Debugging of Three-phase Asynchronous Motor Step-down Starting Circuit

3.2.1 Task Description

Three-phase asynchronous motors have a large starting current that directly affects the power supply transformer, and three-phase squirrel-cage asynchronous motors with larger capacities generally use step-down starting. The step-down start is to reduce the power supply voltage properly and then add it to the stator winding of the motor to start. When the motor starts or is about to end, the motor voltage is restored to the rated value.

The purpose of the reduced voltage start is to reduce the starting current, but the starting torque of the motor will also be reduced. Therefore, the reduced voltage start is only suitable for starting under no load or light load. Common methods for voltage-reduced starting include: stator winding string resistance (or reactor) voltage-reduced starting, $Y-\triangle$ voltage-reduced starting, autotransformer voltage-reduced starting and extended side triangle voltage-reduced starting, etc. This task mainly introduces the realization of the series resistance of the motor and the $Y-\triangle$ step-down start through the contactor, time relay and button.

3.2.2 Task Target

(1) Understand the Y-connection and $\triangle$-connection of the motor.

(2) Understand the meaning, function and method of string resistance buck start and $Y-\triangle$ buck start.

(3) Learn to implement various methods and precautions for buck starting.

3.2.3 Task Analysis

Three-phase asynchronous motors use full-voltage starting (direct starting), and the control circuit is simple; but when the motor capacity is large, direct starting leads to a large starting current, which may reduce the grid voltage and affect the normal operation of other appliances. Full pressure starting is allowed, but reduced voltage starting should be used. Voltage-reduced starting refers to the use of starting equipment to appropriately reduce the voltage and add it to the stator winding of the motor to start. After the motor starts to run, the voltage is restored to the normal operation of the rated value, because the current decreases as the voltage decreases. Therefore, the step-down starting achieves the purpose of reducing the starting current.

This task is divided into four sub-tasks:

(1) Three-phase asynchronous motor stator string resistance step-down start manual control line installation and commissioning;

(2) Three-phase asynchronous motor stator string resistance step-down start automatic control

line installation and commissioning;

(3) Time Relay control three-phase asynchronous motor $Y-\triangle$ start line installation and debugging;

(4) Contactor control three-phase asynchronous motor $Y-\triangle$ start line installation and debugging.

3.2.4 Task-related Knowledge

3.2.4.1 Three-phase Asynchronous Motor Stator String Resistance Step-down Starting Manual Control Circuit

When the motor starts, a resistor is inserted in the three-phase stator circuit, which reduces the voltage of the motor stator winding and limits the starting current. When the motor speed rises to a certain value. Cut off the resistance to make the motor run stably at the rated voltage.

Figure 3-5 is the manual control circuit of the stator string resistance step-down start. Its working process is as follows: press the start button SB_1, the coil of the contactor KM_1 is energized, the self-locking contact and the main contact of the contactor KM_1 are closed, and the motor string resistance start. While the coil of the contactor KM_1 is energized, after a certain period of time, when the motor is started or is about to end, press SB_2, the coil of the contactor KM_2 is energized, the main contact of the contactor KM_2 is closed, and the self-locking Connect the resistor to cut off, the motor is connected to the normal voltage, and enter the normal and stable operation of the person.

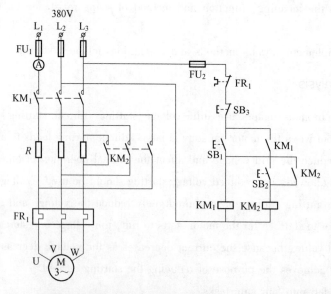

Figure 3-5 Three-phase asynchronous motor stator string resistance step-down starting manual control circuit

Although the stator string resistor step-down starting reduces the starting current, the starting torque is also reduced. This starting method is only suitable for no-load or light-load starting.

3.2.4.2 Automatic Control Circuit of Three-phase Asynchronous Motor Stator String Resistance Step-down Starting

As shown in Figure 3-6, close the power switch QF, press the start button SB_1, the contactor KM_1 and the coil of the time relay KT are energized at the same time, the KM_1 main contact is closed, because the KM_2 coil has a time relay KT in series When the moving contact is closed, it can not be closed. At this time, a resistance R is connected in series in the stator winding of the motor, and the voltage is started to start. The speed of the motor gradually increases. When the time relay KT reaches the preset time, its delay closes. The moving contact is closed, KM_2 is closed, the main contact is closed, the starting resistor R is shorted, and the motor is running at full voltage under the rated voltage. The delay time of KT is usually 4 to 8 seconds.

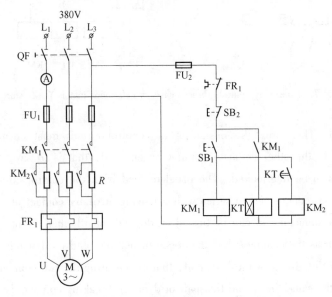

Figure 3-6　Automatic control circuit of three-phase asynchronous motor stator string resistance step-down starting

3.2.4.3 Time Relay Controls Three-phase Asynchronous Motor Y-△ Start Circuit

Motors with a rated connection of delta connection and a larger capacity can use Y-△ step-down starting. When the motor starts, the stator windings are connected in a Y shape, and the voltage drop of each phase winding is $1/\sqrt{3}$ of that in the delta connection. When the speed increases to a certain value, it is changed to △ connection until stable operation. The Y-△ step-down starting control circuit is shown in Figure 3-7.

It can be seen from the figure that there are three sets of main contacts in the main circuit. The main contacts of the contactors KM_2 and KM_3 must not be closed at the same time, because the switch QF closes the power supply, and after the main contact of the contactor KM_1 is closed, the contactors KM_2 and KM_3 If they are closed at the same time, it means that the power supply will

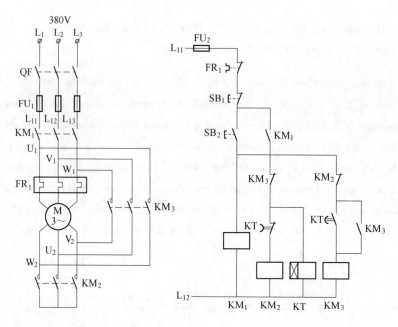

Figure 3-7 Time relay controls three-phase asynchronous motor $Y-\triangle$ start circuit

be short-circuited. Therefore, the design of the control circuit must ensure that when one contactor is engaged, the other contactor cannot be engaged, that is to say, the two contactors KM_2 and KM_3 need to be interlocked. The usual method is in the control circuit. The contactor KM_2 and KM_3 coils are connected in series with a moving auxiliary contact of each other. In this way, whether each contactor coil can be turned on depends on whether the other contactor is in a released state, such as the contactor KM_2 has been turned on, and its dynamic breaking auxiliary contact disconnects the circuit of the KM_3 coil, thereby ensuring the two contactors KM_2 and KM_3 will not pull in at the same time, and this pair of dynamic breaking contacts is called an interlocking contact.

The working principle of the $Y-\triangle$ step-down start control circuit controlled by the time relay is as follows: close the power switch QF and press the start button SB_2. At this time, the contactor KM_1, KM_2, the time relay KT coil is energized, and the contactor KM_1 main contact And self-locking contacts are closed. The main contact of KM_2 is closed and the interlocking contact of KM_2 is disconnected. The motor is started according to the Y-connection. After the set delay time, the dynamic closing contact of the time relay KT is closed and the dynamic breaking contact is opened, so that the contactor The KM_2 coil is powered off, the main contact of the contactor KM_2 is disconnected, the motor is temporarily powered off, and the interlocking contact of the contactor KM_2 is closed. The contactor KM_3 coil is energized, the contactor KM_3 main contact and the self-locking contact are closed, the motor is changed to $\triangle$ connection, and then enters stable operation, at the same time the contactor KM_3 interlocking contact is opened, so that the time relay KT coil is powered off.

3.2.4.4 Contactor Control Three-phase Asynchronous Motor Y-△ Start Circuit

The contactor control three-phase asynchronous motor Y-△ start circuit is shown in Figure 3-8.

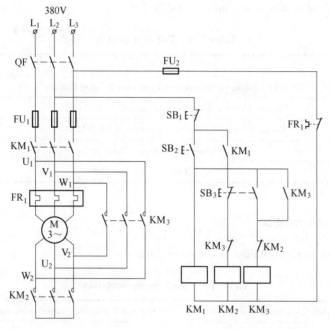

Figure 3-8 Contactor control three-phase asynchronous motor Y-△ start circuit

Circuit control actions are as follows:

Y start: Press SB_2, KM_1 coil is energized, the main contact is closed, and the normally open auxiliary contact is closed to achieve self-locking. While the KM_1 coil is energized, the main contact of KM_2 is closed, and at the same time the normally closed auxiliary contact of KM_2 is opened, which disconnects the path of the KM_3 coil and realizes the interlock. Because the main contacts of KM_1 and KM_2 are closed at the same time, the Y-connected start of the motor is realized.

△ connection operation: After a certain period of time, when the motor starts or is about to end, press SB_3, the SB_3 normally closed contact breaks, the KM_2 coil is de-energized, the KM_2 main contact is reset, and the motor winding seal Y is released. At the same time, the KM_2 normally closed auxiliary contact is reset and the interlock is released. While pressing SB_3, the SB_3 normally open contact is closed, the KM_3 coil is energized, its main contact is closed, and the motor winding is connected to run. At the same time, the normally closed contact of KM_3 is disconnected, disconnecting the coil path of KM_2 and achieving interlocking.

Stop: Press SB_1, the control circuit is powered off, each contactor is released at the same time, and the motor stops.

The starting current of the Y connection is only three-thirds of the delta connection, which limits the starting current, but the starting torque of the Y connection is one third of the △ connection, so the Y-△ start is only applicable to no load Or light load start.

3.2.5 Bill of Materials

The table of the Bill of materials is shown in Table 3-6.

Table 3-6　Bill of materials

No.	Name	Number
1	XK-SX2C Advanced Maintenance Electrician Training Platform	1
2	Repair Electrician Training Components (XKDT11)	1
3	Repair Electrician Training Components (XKDT12A)	1
4	Three-phase Asynchronous Motor	1
5	Jumper	Several

3.2.6 Task Implementation

The table of the task implementation is shown in Table 3-7.

Table 3-7　Task implementation

Three-phase asynchronous motor step-down start line installation and commissioning task book					
1. Complete the required device inventory and complete the following table as required					
No.	Device	Number	Parameters	Measurement	Remarks
2. Working principle description					

Continued Table 3-7

Three-phase asynchronous motor step-down start line installation and commissioning task book	
3. Design and drawing of circuit wiring diagram	
4. Operation process record	
5. Recording process of power-on test drive	
6. Task summary	

3.2.7 Task Evaluation

The table of the task evaluation is shown in Table 3-8.

Table 3-8　Task evaluation

Three-phase asynchronous motor $Y-\triangle$ step-down start line installation and commissioning score table

Device inventory and measurement (10 points)

No.	Key inspection	Grading	Total score	Score	Remarks
1	Device inventory	1 point for one error	5		
2	Device measurement	1 point for one error	5		
	Subtotal				

Circuit diagram design (10 points)

No.	Key inspection	Grading	Total score	Score	Remarks
1	Main circuit design	1 point for one error	5		
2	Control circuit design	1 point for one error	5		
	Subtotal				

Working principle description (10 points)

No.	Key inspection	Grading	Total score	Score	Remarks
1	Working principle of main circuit	Complete discretion	5		
2	Working principle of control circuit	Complete discretion	5		
	Subtotal				

Line construction (30 points)

No.	Key inspection	Grading	Total score	Score	Remarks
1	Whether the wire is handled correctly	1 point for one error	10		
2	Whether the wire is firmly installed	1 point for one error	10		
3	whether the line correct	1 point for one error	10		
	Subtotal				

Power on test (30 points)

No.	Key inspection	Grading	Total score	Score	Remarks
1	Y start function	Whether the function is realized	10		
2	$\triangle$ Connect to manual start function	Whether the function is realized	10		
3	$\triangle$ Connect auto start function	Whether the function is realized	10		
	Subtotal				

Professional accomplishment (10 points)

No.	Key inspection	Grading	Total score	Score	Remarks
1	Power on operation	5 point for one error			
2	Standard operation	1 point for one error			
3	Team work	Discretion			
4	Work place tidy	Discretion			
	Total				

3.2.8 Troubleshooting

(1) The contactor controls the three-phase asynchronous motor $Y-\triangle$ Start line Y. The starting process is normal, but after pressing SB_3, the motor makes an abnormal sound, and the speed drops sharply.

Analysis phenomenon: The switching action of the contactor is normal, indicating that the control circuit wiring is correct. The problem occurs after connecting the motor. From the analysis of the fault phenomenon, it is likely that the main circuit wiring of the motor is incorrect. When the circuit is switched from Y to $\triangle$, the phase sequence of the power supplied to the motor changes, and the motor suddenly changes from normal start It became the reverse sequence power supply braking, and the powerful reverse braking current caused the motor speed to drop sharply and abnormal sound.

Troubleshooting: Check the wiring sequence of the main circuit contactor and the motor terminal.

(2) The line no-load test works normally. When the motor is connected to the test drive, the motor will move together and the motor will make an abnormal sound. The rotor will vibrate left and right. Press SB_1 to stop immediately. When stopping, there is a strong arc phenomenon in the arc extinguishing cover of KM_2 and KM_3.

Analysis phenomenon: The contactor switching action is normal during the no-load test, indicating that the control circuit wiring is correct. The problem occurs after the motor is connected, and the analysis of the fault phenomenon is caused by the lack of phase of the motor. When the motor starts at Y, one phase winding is connected to the circuit. The motor causes single-phase starting. Because the missing phase winding cannot form a rotating magnetic field, the rotation of the motor's rotating shaft is unstable and vibrates left and right.

Troubleshooting: Check whether the contactor contacts are closed properly, and whether the contactor and motor terminal wiring are tight.

(3) When the time relay controls the three-phase asynchronous motor $Y-\triangle$ start line no-load test, as soon as the start button SB_2 is pressed, KM_2 and KM_3 will switch on and off and cannot be switched on.

Analyze the phenomenon: Repeat the switching action by starting KM_2 and KM_3, indicating that there is no delay action of the time relay. As soon as the SB_2 start button is pressed, the time relay coil is electrically attracted, and the contact also acts immediately, causing the mutual switching of KM_2 and KM_3, which cannot be Start normally.

Troubleshooting: The problem occurs at the contact of the time relay. Check the wiring of the time relay, and find that the contact of the time relay is used incorrectly, and it is connected to the instantaneous contact of the time relay, so it will act as soon as the contact is energized, and the line will be connected to the delay contact of the time relay for troubleshooting. Time relays often have a pair of delayed action contacts and a pair of instantaneous action contacts. Check the use requirements of the time relay contacts before wiring.

The table of the troubleshooting record is shown in Table 3-9.

Table 3-9 Troubleshooting record

	Troubleshooting record	
Fault 1	Fault phenomenon	
	Cause of fault	
	Trouble shooting process	
Fault 2	Fault phenomenon	
	Cause of fault	
	Trouble shooting process	
Fault 3	Fault phenomenon	
	Cause of fault	
	Trouble shooting process	
Fault 4	Fault phenomenon	
	Cause of fault	
	Trouble shooting process	

3.2.9 Task Development

In this task, the three-phase asynchronous motor $\Upsilon-\triangle$ step-down control is used to start. In some occasions, the step-down starting can be achieved through the triangle connection of the three-phase asynchronous motor, and try to complete the design.

Mission accomplished:

(1) Complete the design of the circuit diagram;

(2) Complete the principle explanation of 'Three-phase asynchronous motor extended-edge triangular buck start control circuit'.

3.2.10 Task Summary

Voltage-reduced starting refers to the use of starting equipment to appropriately reduce the voltage and add it to the stator winding of the motor to start. After the motor starts to run, the voltage is restored to the normal operation of the rated value, because the current decreases as the voltage decreases. Therefore, the step-down starting achieves the purpose of reducing the starting current. But at the same time, because the torque of the motor is proportional to the square of the voltage, the buck start will also cause the motor's starting torque to be greatly reduced. Therefore, the step-down start needs to be carried out under no load or light load.

During the implementation of the task, the power supply should be disconnected first, and the experimental wiring should be carried out according to the figure. First, power on the control loop to see whether the contactor is operating correctly. After the control loop is debugged correctly, then power on the main loop to see if the motor is running correctly. During the implementation process, it is not possible to carry out the wiring operation with power on. The control loop should be debugged first and then the main loop.

Project 4 Troubleshooting of Electrical Control Circuits of Typical Machine Tools

Task 4.1 Fault Maintenance of Electrical Control Circuit of CA6140 Lathe

4.1.1 Task Description

Machine, also known as machine tools or machine tool, the function of machine tool is manufacturing machine, they are mainly used in industrial machinery and equipment, transportation equipment, primary metal products and electrical equipment and machinery manufacturing equipment, the basis of widely used in various industrial applications, including metal cutting machine tool according to processing methods can be divided into the lathe, drilling machine, boring machine, grinding machine, etc. This task is to master the CA6140 lathe electrical control line fault elimination method.

4.1.2 Task Target

(1) Understand the function, structure and movement form of the lathe.
(2) Familiar with the composition and working principle of electrical control circuit of CA6140 lathe.
(3) Master the fault elimination method of CA6140 lathe electrical control line.

4.1.3 Task Analysis

Lathe is a widely used metal cutting machine tool, which is mainly used for machining various rotary surfaces and end faces of rotary bodies. Such as turning inside and outside cylindrical surface, cone surface, ring groove and forming rotary surface, turning end face and all kinds of common threads, equipped with technology and equipment can also process all kinds of special surface. Its characteristic is that the turning tool is relatively fixed and the workpiece is rotating at high speed. This task is divided into four sub-tasks:

(1) to understand the main structure of the lathe, the form of movement and the characteristics of electric drive, control requirements.
(2) Read the electrical schematic diagram of CA6140 ordinary lathe.
(3) To install and debug the electrical control circuit of CA6140 ordinary lathe.
(4) Carry out the analysis and maintenance of common electrical faults of CA6140 common lathe.

4.1.4 Task-related Knowledge

4.1.4.1 Main Structure and Movement Form of Lathe

CA6140 lathe model significance:

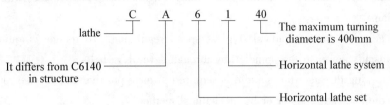

The appearance and main structure of CA6140 ordinary lathe are shown in Figure 4-1.

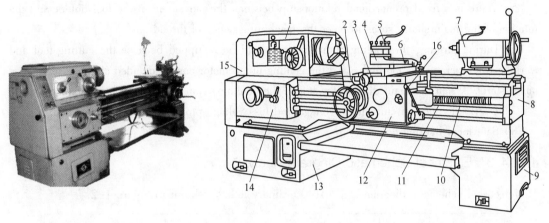

Figure 4-1 Appearance and main structure of lathe

1—Spindle box; 2—Vertical slide plate; 3—Horizontal slide plate; 4—Turntable; 5—Square tool rest; 6—Small slide plate; 7—Tailstock; 8—Bed body; 9—Right bed base; 10—Light bar; 11—Lead screw; 12—Slide box; 13—Left bed box; 14—Feed box; 15—Wheel rack; 16—Control handle

CA6140 ordinary lathe is mainly composed of bed, spindle box, feed box, sliding board box, knife rest, lead screw, light bar, tail frame and other parts.

The cutting motion of a lathe includes the main motion of a chuck or a center that drives the workpiece to rotate and the linear feed motion of a slide plate that drives the tool holder and the tool. When the lathe works, most of the power is consumed in the spindle motion. Turning speed refers to the relative speed of workpiece and tool contact point; According to the workpiece material property, turning tool material and geometry, workpiece diameter, processing mode and cooling conditions, the spindle has different cutting speed. Spindle speed change is achieved by the spindle motor through the V-belt transmission to the spindle gearbox. The CA6140 lathe has 24 kinds of spindle forward speed (10~1400r/min) and 12 kinds of reverse speed (14~1580r/min).

The feed motion of a lathe is the linear motion driven by the tool holder. The sliding plate box transfers the rotation of the lead screw or the light bar to the tool holder part, changes the position of the handle outside the sliding plate box, and makes the turning tool do longitudinal or transverse feed through the tool holder part.

Auxiliary movement of the lathe is all necessary movement except cutting movement in the lathe, such as longitudinal movement of the tailstock, clamping and loosening of the workpiece.

4.1.4.2 Characteristics and Control Requirements of Electric Drive

(1) The main drive motor is generally three-phase cage asynchronous motor, without electrical speed regulation.

(2) Adopt gear box for mechanical stepless speed regulation. To reduce vibration, the main drag motor transmits power to the spindle box through several V-belts.

(3) When turning threads, the spindle is required to have positive and reverse rotation, which is realized by the main drag motor or by mechanical method.

(4) The start and stop of the main drag motor shall be operated by buttons.

(5) There is a fixed proportional relationship between the movement of the tool holder and the rotation of the main shaft, so as to meet the machining needs of threads.

(6) During turning, the cooling pump motor should be equipped because the cutting tool and workpiece temperature is too high, and the cooling pump motor can be decided after the main drag motor starts. When the main drag motor stops, the cooling pump should stop immediately.

(7) Overload, short circuit, undervoltage and loss voltage protection must be provided.

(8) Safe local lighting device.

4.1.4.3 Electrical Schematic Diagram Analysis

The electrical schematic diagram of CA6140 ordinary lathe is shown in Figure 4-2.

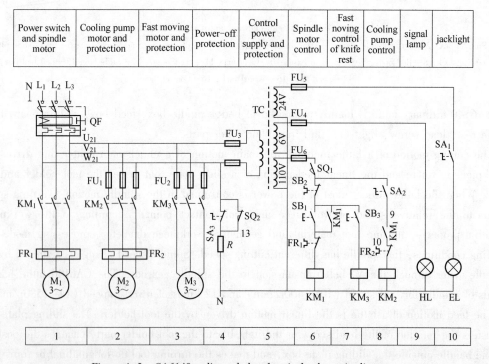

Figure 4-2 CA6140 ordinary lathe electrical schematic diagram

(1) Main circuit analysis. The power supply is introduced by leakage protection circuit breaker QF. The operation and stop of spindle motor M_1 is controlled by the connection and disconnection of the three normally open main contacts of contactor KM_1. The capacity of motor M_1 is not large, so direct starting is adopted. The operation and stop of the cooling pump motor M_2 is controlled by the three normally open main contacts of the contactor KM_2. The operation and stopping of M_3, a fast moving motor, is controlled by the three normally open main contacts of the contactor KM_3. The power end of leakage protection circuit breaker QF shall be connected with fuse for short circuit protection. Both the Cooling pump motor M_2 and the fast-moving motor M_3 have very small capacities and are protected by fuses FU_1 and FU_2 for short circuit protection. Thermal relays FR_1 and FR_2 are used for overload protection of M_1 and M_2 respectively, and their thermal elements are connected to their respective main circuits. The fast moving motor M_3 works in a short time, so there is no overload protection.

(2) Control circuit analysis. The control circuit is supplied with 110V AC voltage, which is obtained by 380V voltage step-down from the control transformer TC. The primary circuit is protected by fuse FU_3 and the secondary circuit is protected by FU_6. First close the leakage protection circuit breaker QF and assume that the normally open contact of the trip switch SQ_1 is closed.

1) Control of the spindle motor: press the starting button SB_1, the contactless KM_1 coil is energized, the KM_1 core pulls in, the three normally open main contacts of KM_1 on the main circuit are closed, and the spindle motor M_1 starts. At the same time, a normally open auxiliary contact of KM_1 is also closed for self-locking to ensure the continuous rotation of spindle motor M_1 after releasing the starting button. Press the stop button SB_2, and the contactor KM_1 will be released due to the coil power failure. Its three normally open main contacts will be disconnected, and the spindle motor M_1 will stop. The normally closed contact of thermal relay FR_1 is connected in series in the circuit of KM_1 coil. When the spindle motor M_1 is overloaded, the normally closed contact of FRl will be disconnected, and KM_1 will be released due to the coil power failure, and the motor M_1 will stop. The circuit has zero voltage protection function. After the power is cut off, the contactor KM_1 is released. When the power voltage returns to normal again, if the start button SB_1 is not pressed, the motor will not start by itself, so as to avoid accidents. The circuit is also protected by undervoltage. When the power supply voltage is too low, the contortor KM_1 will be released automatically due to insufficient electromagnetic suction, and the motor M_1 will stop automatically to avoid the motor M_1 burning out due to excessive current when the voltage is under.

2) cooling pump motor when the spindle motor, the control of normally open auxiliary contact of KM_1 (9-10) closed, if you need cooling fluid at this moment, the rotating switch SA_2, make its closure, the contactor KM_2 coil electricity, iron absorption, main circuit, the three normally open at KM_2 main contact closure, cooling pump motor starting operation, providing cutting coolant. When the spindle motor stops, the contortor KM_1 is released, its normally open contact (9-10) is disconnected, and the cooling pump motor M_2 stops at the same time. It can be seen that the cooling pump motor M_2 can only be started when the spindle motor M_1 is started, and there is an interlock between them. The normally closed contact of thermal relay FR_2 is connected in series in the

circuit of KM_2 coil, so when the cooling pump motor is overloaded, the normally closed contact of FR_2 will be disconnected, and the contactor KM_2 will be released due to the coil power failure, and the motor M_2 will be stopped to achieve overload protection. Contactor KM_2 also has undervoltage protection for cooling pump motor.

3) Control of fast moving motors Fast moving is the inching control circuit. Press the button SB_3, the contactor KM_3 coil is electrified, the iron core pulls together, and the three normally open main contacts of KM_3 in the main circuit are closed, the M_3 motor is moved quickly, and the knife rest is moved quickly. Release the button and KM_3 will stop. The direction of rapid movement is controlled by turning the cross handle mounted on the slide box in the desired direction.

In addition to the above interlock between spindle motor M_1 and cooling pump motor M_2, there are other interlocks.

4) The contact (W_{21}-13) of the interlock protection key power switch SA_3 is connected in parallel with the normally closed contact of the travel switch SQ_2 in series with the leak detection resistance R. The condition for detecting the current of leakage resistance is that the key power switch is rotated to the SA_3 disconnect position with the key and the cover of the electrical box is covered. At this time, the cover is pressed down on the travel switch SQ_2 and the normally closed contact is disconnected. Only in this case, leakage resistance is not switched on, leakage protection switch QF can be closed, to ensure safety. SQ_1 is the safety travel switch of the rack. When the wheel rack cover is installed, SQ_1's normally open contact is closed so that the control circuit can be powered and the motor can be started.

(3) Analysis of lighting circuit and signal indicating circuit.

Lighting circuit adopts 24V AC voltage. The lighting routing switch SA_1 is composed of light bulb EL. The other end of the lamp EL must be grounded to prevent electrical shock that could occur in the event of a short circuit between the original and secondary winding of the lighting transformer. Fuse FU_5 is the short circuit protection of lighting circuit. The signal indicating circuit adopts 6V AC voltage, indicating that the bulb HL is connected to the 6V coil of the control transformer TC secondary, and the indicator light is on, indicating that the control circuit has electricity. Fuse FU_4 is short circuit protection of signal indicating circuit.

4.1.4.4 Electrical fault maintenance of CA6140 ordinary lathe

(1) Analysis and maintenance of common electrical faults. When it is necessary to open the switchboard niche door for live maintenance, pull out the transmission lever of SQ_2 switch and the circuit breaker QF can still be closed. SQ_2 restores the protective effect after closing the niche door.

1) Spindle motor M_1 cannot be started and can be repaired in the following steps:

Check whether the contactor KM_1 pulls in. If the contactor KM_1 pulls in, the fault must occur in the power circuit and the main circuit. You can follow the following steps:

① Close the circuit breaker QF and measure the voltage between U_{21}, V_{21} and W_{21} at the receiving end of the contactor with a multimeter. If the voltage is 380V, the power circuit is normal.

When there is no voltage between U_{21} and W_{21}, then measure the voltage between L_1, L_2 and L_3. If there is no voltage, it indicates power failure. If there is voltage, the circuit breaker QF has bad contact or broken connection.

Repair measures: to find out the cause of damage, replace the circuit breaker and connection wire of the same specification and model.

② Disconnect QF of the circuit breaker and measure the resistance between the main contacts of contortionists KM_1 with the multimeter resistor R_{X1}. If the resistance is small and equal, the measured circuit is normal; Otherwise; Check FR_1, Motor M_1 and the connection between them in turn.

Repair measures: to find out the cause of damage, repair or replace the thermal relay FR_1 and motor M_1 of the same specification and type and the connection wires between them.

③ Check whether the main contact of the contactor KM_1 is in good condition. If the contact is defective or burnt, change the moving or static contact or the contactor of the same specification.

④ Check whether the mechanical parts of the motor are in good condition. If the internal bearings of the motor are damaged, the bearings should be replaced. If there is any problem with external machinery, you can cooperate with the mechanic to repair it.

If contactor KM_1 does not suck, repair can follow the following steps:

First, check whether KM_3 is suction. If the suction indicates that part of the common control circuit of KM_1 and KM_3 is normal, the fault range is in the coil branch of KM_1. If KM_3 does not suck, it is necessary to check whether the lamp and signal lamp are on. If the lamp and signal lamp are on, it indicates that the fault range is on the control circuit. If the lamp HL and EL are not on, it indicates that there is a fault in the power supply, but there is also a fault in the control circuit.

2) Spindle motor M_1 cannot be self-locked after starting; When the start button SB_1 is pressed, the spindle motor will start, but after the release of SB_1, M_1 will also stop. The cause of this failure is a poor self-locking contact of the contactor KM_1 or a loose connection wire.

3) The failure of spindle motor M_1 to stop is mainly caused by the fusion welding of the main contact of contactless KM_1; Stop button SB_2 breakdown or SB_2 two point connection wire in the line short circuit; The surface of the contactor core is firmly covered with dirt. The following methods can be used to determine which cause the motor M_1 cannot be stopped: if QF is disconnected, the contactor KM_1 is released, indicating that the fault is SB_2 breakdown or short wire connection; If the contactor is released after a period of time, the fault is that the surface of the core is stuck to dirt. If QF is disconnected and contactor KM_1 is not released, the primary contact will fuse. Take corresponding measures to repair according to specific faults.

4) The main reason for the failure of spindle motor to stop suddenly in operation is the action of thermal relay FR_1. After such a fault occurs, it is necessary to find out the cause of the thermal relay FR_1 operation and reset it only after removing it. The cause of action of thermal relay FR_1 may be: three-phase power supply voltage imbalance; The power supply voltage is too low for a long time; Heavy load and poor contact of M_1 connection wire, etc.

5) The tool holder fast moving motor cannot be started. First check whether the FU_2 fuse is fusible; Secondly, check whether the contact of KM_3 contact is good; If there is no abnormality or when SB_3 is pressed, the contactor KM_3 does not snap, then the fault must be in the control circuit. At this time, check whether the coil of the inking button SB_3 and contactor KM_3 has the phenomenon of circuit break.

(2) Maintenance procedures and technological requirements.

1) Operate the lathe under the guidance of the operator, and understand various working conditions and operation methods of the lathe.

2) Under the guidance of the teacher, be familiar with the distribution and wiring of the electrical components of the lathe by referring to the electrical location map and machine tool wiring diagram.

3) Artificial setting of natural fault points on the electrical control line of CA6140 lathe. The following points should be noted in the fault setting:

① The fault must be set to simulate the lathe in use, due to the influence of external factors caused by the natural fault.

② Do not set to change the line or replace electrical components due to man-made causes of non-natural failure.

③ For lines with more than one fault point, the fault phenomenon should not be covered up. If faults cover each other, there should be obvious inspection order as required.

④ Should try to set up not easy to cause personal or equipment accidents fault points.

(3) Matters needing attention.

1) Familiar with the basic links and control requirements of the electrical control circuit of CA6140 lathe, carefully observe the teacher's demonstration maintenance.

2) The tools and instruments used in maintenance shall meet the use requirements.

3) When troubleshooting, the fault point must be repaired, but component substitution is not allowed.

4) During maintenance, it is strictly prohibited to expand the fault scope or generate new faults.

5) During live maintenance, the instructor must be in charge to ensure safety.

4.1.5 Bill of Materials

The table of the Bill of materials is shown in Table 4-1.

Table 4-1 Bill of materials

No.	Name	Number
1	XK-SX2C Advanced Maintenance Electrician Training Platform	1
2	Repair Electrician Training Components (XKDT11)	1
3	Repair Electrician Training Components (XKDT12A)	1
4	Three-phase Asynchronous Motor	1
5	Jumper	Several

4.1.6 Task Implementation

The table of the task implementation is shown in Table 4-2.

Table 4-2 Task implementation

CA6140 type general lathe electrical Circuit installation, debugging tasks					
1. Complete the required device inventory and complete the following table as required					
No.	Device	Number	Parameters	Measurement	Remarks
2. Working principle description					
3. Design and drawing of circuit wiring diagram					

Continued Table 4-2

CA6140 type general lathe electrical Circuit installation, debugging tasks
4. Operation process record
5. Recording process of power-on test drive
6. Task summary

4.1.7 Task Evaluation

The table of the task evaluation is shown in Table 4-3.

Table 4-3 Task evaluation

CA6140 type general lathe electrical circuit installation, debugging rating table					
Device inventory and measurement (10 points)					
No.	Key inspection	Grading	Total score	Score	Remarks
1	Device inventory	1 point for one error	5		
2	Device measurement	1 point for one error	5		
		Subtotal			
Circuit diagram design (10 points)					
No.	Key inspection	Grading	Total score	Score	Remarks
1	Main circuit design	1 point for one error	5		
2	Control circuit design	1 point for one error	5		
		Subtotal			
Working principle description (10 points)					
No.	Key inspection	Grading	Total score	Score	Remarks
1	Working principle of main circuit	Complete discretion	5		
2	Working principle of control circuit	Complete discretion	5		
		Subtotal			

Continued Table 4-3

CA6140 type general lathe electrical circuit installation, debugging rating table

Line construction (30 points)

No.	Key inspection	Grading	Total score	Score	Remarks
1	Whether the wire is handled correctly	1 point for one error	10		
2	Whether the wire is firmly installed	1 point for one error	10		
3	whether the line correct	1 point for one error	10		
		Subtotal			

Power on test (30 points)

No.	Key inspection	Grading	Total score	Score	Remarks
1	Spindle control function	Whether the function is realized	10		
2	Cooling pump control function	Whether the function is realized	10		
3	Fast moving control function	Whether the function is realized	5		
4	Lighting, signal indicating control	Whether the function is realized	5		
		Subtotal			

Professional accomplishment (10 points)

No.	Key inspection	Grading	Total score	Score	Remarks
1	Power on operation	5 point for one error			
2	Standard operation	1 point for one error			
3	Team work	Discretion			
4	Work place tidy	Discretion			
		Total			

4.1.8 Troubleshooting

The table of the troubleshooting record is shown in Table 4-4.

Table 4-4 Troubleshooting record

	Troubleshooting record	
Fault 1	Fault phenomenon	
	Cause of fault	
	Trouble shooting process	
Fault 2	Fault phenomenon	
	Cause of fault	
	Trouble shooting process	

Continued Table 4-4

		Troubleshooting record
Fault 3	Fault phenomenon	
	Cause of fault	
	Trouble shooting process	
Fault 4	Fault phenomenon	
	Cause of fault	
	Trouble shooting process	

4.1.9 Task Development

General method of electrical fault repair of machine tools:

(1) Failure investigation before maintenance When electrical faults occur in industrial machinery, do not start maintenance blindly. Before maintenance, ask, see, listen and feel to understand the operation situation before and after the failure and the abnormal phenomenon after the failure, so as to judge the fault location according to the fault phenomenon, and then accurately eliminate the fault.

Q: Ask the operator about the operation status of the circuit and equipment before and after the failure and the symptoms after the failure; If the fault is frequent or occasional; Whether there are signs of noise, smoke, sparks, abnormal vibrations, etc. ; Whether the cutting force is too large and frequently start, stop and brake before the fault occurs; There is no maintenance, repair or wiring changes.

See: check whether there are obvious appearance signs before the fault occurs, such as various signals; Fuses with indicating devices; Trip action of protective electrical appliance; The wiring comes off; Contact ablation or fusion welding; The coil overheats and burns out.

Listen: On the premise that the line can still run and the fault scope is not extended and the equipment is not damaged, the test can be powered on and listen to whether the sound of electric motors, contactors and relays is normal.

Touch: check the motor, transformer, solenoid coil and fuse as soon as possible after the power is just cut off to see if there is overheating.

(2) When using logical analysis method to determine and narrow the scope of faults for simple electrical control lines, check each electrical component and each wire one by one, and generally find fault points quickly. But for complex circuits, there are often hundreds of components, thousands of connections, if you take the method of checking one by one; Not only does it take a lot of time, but it is also easy to miss. In this case, according to the circuit diagram, the logic analysis method can be used to make a specific analysis of the fault phenomenon, draw out a suspicious range, improve the pertinence of maintenance, can receive accurate and fast results. Analysis circuit, usually first, from the perspective of the main circuit, realize the industrial machinery using a few moving parts and electric motor drive, what are related to each motor electrical components,

what control adopted, and then according to the electric main circuit of the electrical components used text symbols, sex and control requirements, find the corresponding control circuit. On this basis, combined with the fault phenomenon and the principle of line work, careful analysis and troubleshooting, can quickly determine the possible scope of the fault.

When the suspicious scope of the fault is large, the fault can be checked in the middle of the scope to determine which part of the fault occurred, so as to narrow the scope of the fault and improve the maintenance speed.

(3) Inspect the appearance of the scope of the fault. After determining the possible scope of the fault, inspect the appearance of the electrical components and connection wires within the scope, such as: fuse fuse; Wire joints loose or fall off; Contacts of contactors and relays fall off or have bad contact, the coil burns and causes the surface insulating paper to burn and discolor, and the burnt insulating varnish flows out; Spring falls off or breaks; The action mechanism of electrical switch is blocked and malfunctioning, etc. , which can clearly indicate the fault point.

(4) using test method further narrowing the scope of the fault by the appearance inspection found no fault point, can according to the fault phenomenon, combined with the circuit diagram analysis the cause of the problem, in expanding the scope of the fault, under the premise of damage electrical and mechanical equipment, direct current test, or remove load (off) from the control box terminal plate electric test, in order to distinguish the fault may be in electrical parts and machinery, and other parts; On the motor or on the control equipment; Is it on the main circuit or the control circuit? In general, check the control circuit first. To be specific, when a button or switch is operated, the contactors and relays in the circuit will work in the prescribed action sequence. If the sequence of action to a certain electrical element, found that the action does not meet the requirements, that means that the electrical element or its related circuits have problems. Then in the circuit to carry out item by item analysis and inspection, generally can find the fault. After the fault of the control circuit is eliminated and restored to normal, the main circuit shall be closed to check the control effect of the control circuit on the main circuit and observe whether there is any abnormality in the working condition of the main circuit.

Attention must be paid to the safety of person and equipment during the electrification test. Must observe the safe operation procedure, may not touch the live part at will, must cut off as far as possible the electric machine main circuit power supply, only in the control circuit under the condition of live inspection; If the motor is needed to run, it should be run under no load to avoid the misoperation and collision of the moving part of the industrial machinery. To temporarily cut off the fault of the main circuit, so as not to expand the fault, and fully estimate in advance of the local line action can occur after the adverse consequences.

(5) Determination of Fault Points by Measuring Method Measurement method is an effective inspection method used to accurately determine fault points in maintenance electrical work. Commonly used testing tools and instruments include calibration lamps, test pens, multimeters, clamp ammeters, megometers, etc. , which are mainly used to measure the relevant parameters such as voltage, resistance and current when the circuit is electrification or power off, to judge the quality

of electrical components, the insulation of equipment and the on-off condition of the circuit. With the development of science and technology, measurement methods are constantly updated. For example, in the thyristor - motor automatic speed regulation system, the fault of the system can be quickly judged by using the oscilloscope to observe the output waveform of the thyristor rectifier device and the pulse waveform of the trigger circuit.

When using measurement method to check the fault point, it is necessary to ensure that all kinds of measuring tools and instruments are in good condition and used correctly. It is also necessary to pay attention to prevent the influence of induction electricity, circuit electricity and other parallel branches, so as to avoid false judgment.

4.1.10 Task Summary

To type CA6140 engine lathe electrical wiring installation, debugging and troubleshooting, for example, to understand common methods of machine tool electrical troubleshooting, through investigation before repair, ask, see, hear, touch, understand the fault before and after operation and after the failure of anomalies, using logical analysis to identify and narrowing the scope of the fault, and through the correct use of all kinds of measuring tools and instruments measuring point of failure, for troubleshooting.

Task 4.2 Z35 Radial Drilling Machine Electric Control Line Fault Maintenance

4.2.1 Task Description

Drilling machine is a versatile machine tool with wide use and also a metal cutting machine tool. It can drill holes, reaming, counterfacing and tapping threads for spare parts. The drilling machine is equipped with process equipment and can also bore holes. Equipped with a universal table can also be drilling, reaming, reaming. This task is to master the fault elimination method of Z35 radial drilling machine control line.

4.2.2 Task Target

(1) Understand the function, structure and movement form of the drilling machine.
(2) Familiar with the composition and working principle of Z35 radial drilling machine control line.
(3) Master the fault elimination method of Z35 radial drilling machine control line.

4.2.3 Task Analysis

A drill press is a machine tool that mainly USES a bit to make holes in the workpiece. The structure of drilling machine is simple, the processing precision is relatively low, the characteristics of drilling machine is the workpiece is fixed, the tool does rotation motion.

This task is divided into four subtasks:

(1) To understand the main structure, movement form, characteristics and control requirements of Z35 radial drilling machine.

(2) Read the electrical schematic diagram of Z35 radial drilling machine.

(3) To install and debug the electrical control circuit of Z35 radial drilling machine.

(4) Common electrical fault analysis and maintenance of Z35 radial drilling machine.

4.2.4 Task-related Knowledge

4.2.4.1 Main Structure and Movement of Radial Drilling Machine

Definition of Z35 radial drilling machine model:

$$\text{Drilling machine} \underset{\uparrow}{Z} \underset{\uparrow}{3} \underset{\uparrow}{5} \begin{matrix} \text{Maximum drilling diameter is 50mm} \\ \text{Radial drilling machine} \end{matrix}$$

The outline and main structure of radial drilling machine are shown in Figure 4-3. An inner column is fixed on the base, and an outer hollow column is arranged on the outside of the inner column. The rocker arm can rotate around the inner column along with the outer column. The rocker arm and the outer pillar cannot rotate relative to each other. The spindle box can be moved horizontally along the guide rail on the rocker arm. Because of these movements, it is easy to adjust the position of the drill bit on the spindle relative to the workpiece to align with the center of the machining hole required for the workpiece. Therefore, some large and heavy porous workpiece can be easily machined on the radial drilling machine without moving the workpiece.

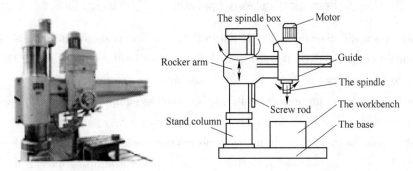

Figure 4-3 Outline and main structure of radial drilling machine

When the workpiece is not big, it can be pressed on the worktable for processing. If the workpiece is large, it can be machined directly on the base. Depending on the height of the workpiece, the rocker arm can be lifted up and down along the outer column with the help of a lead screw. Before lifting, the rocker arm should be automatically released and lifted again. When the desired lifting position is reached, the rocker arm is automatically clamped to the column. The release and clamping of Z35 radial drilling machine depends on the mechanical mechanism automatically. The rocker arm and the outer column rotate around the inner column by manpower, but the outer column must be released first. The horizontal movement of the spindle box along the upper guide rail of the rocker arm is also manual. The spindle box must also be released first. When machining is required, the outer column shall be clamped to the inner column and the headstock shall be clamped to the rocker arm. In this way, the spindle position will not move and the cutter will not vibrate during machining. Therefore, the movement of radial drilling machine includes: the main movement is the spindle driving the rotary movement of the drill bit; Feed movement is the movement of the bit up and down; Auxiliary movements include horizontal movement of the spindle box along the rocker arm, up and down movement of the rocker arm along the outer column, and rotation of the rocker arm with the outer column relative to the inner column.

4.2.4.2 Characteristics and Control Requirements of Electric Drive

(1) The drilling machine has many moving parts, which are driven by multiple motors.

(2) Drilling machine is sometimes used for tapping, so the main shaft can be positive and negative rotation. The positive and negative rotation of the main shaft is generally achieved by the positive and negative friction clutch. The Z35 radial drill press is controlled by the handle push pressure spring and the lever to control the positive and negative rotation friction clutch. Therefore, for the spindle motor, only one direction rotation is required. The main shaft generally adopts three-phase cage asynchronous motor to drag, with speed change mechanism to adjust the spindle speed and feed, spindle speed change and feed change mechanism are in the main shaft box.

(3) The tightness of the outer column and the spindle box is carried out simultaneously by the hydraulic push elastic mechanism.

4.2.4.3 Electric Schematic Diagram Analysis of Z35 Radial Drilling Machine

The electrical schematic diagram of Z35 radial drilling machine is shown in Figure 4-4.

(1) The main circuit analysis power supply is introduced by the transfer switch QS to prepare the machine tool for additional work. The whole machine is protected by a fuse.

There are four motors in the main circuit. M_1 is a cooling pump motor, which provides coolant to the workpiece and is directly controlled by the transfer switch SA_2. M_2 is the spindle motor, and its start and stop is controlled by the normally open main contact of the contortor KM_1, and the thermal relay FR ACTS as overload protection. M_3 is a rocker arm lifting motor, and its positive and negative rotation is controlled by the normally open main contact of the contactor KM_2 and KM_3. M_4 is a column relaxation and clamping motor, and its positive and negative rotation is controlled by the normally open main contact of contactor KM_4 and KM_5. Both the motor M_3 and the motor M_4 operate in a short time, so there is no overload protection. M_3, M_4 and the control circuit share fuse FU_2 for short circuit protection. Since the outer column and rocker arm rotate around the inner column, all power sources except the cooling pump are introduced through bus A.

(2) Analysis of control circuit for the sake of safety, the power supply voltage used in the control circuit is AC 127V, which is obtained by reducing the AC voltage of 380V by transformer TC.

Z35 radial drilling machine control circuit using the cross switch SA_1 operation, it has the ad-

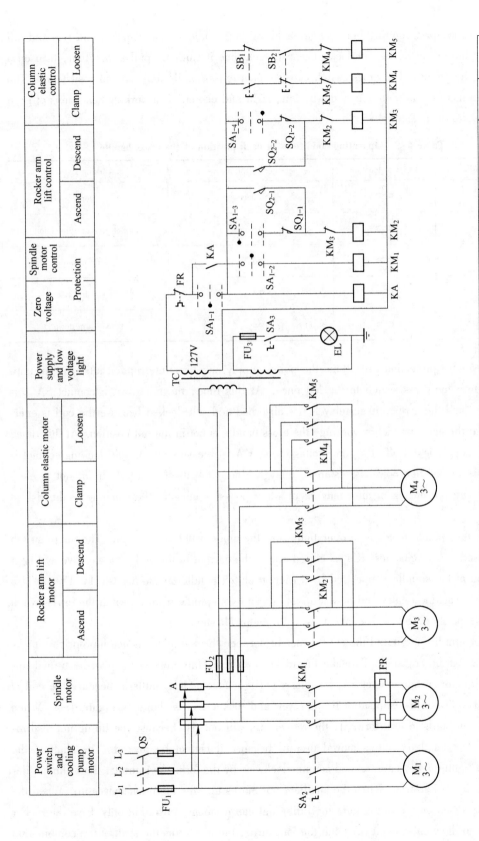

Figure 4-4　Z35 radial drilling machine electrical schematic diagram

vantages of centralized control, but also can achieve the spindle stop and rocker arm rise and fall between the movement of the interlock, because each time it can only pull a direction, turn on a direction of the circuit. The cross switch consists of a cross handle and four microswitches. The cross handle has five positions: up, down, left, right and center. The working conditions of each position are shown in Table 4-5.

Table 4-5 Operating conditions of each position of the cross handle

Handle position	Physical location	Get accessed to the microswitch	Working state
Middle		No access	Stop
Left		SA_{1-1}	Loss of pressure protection
Right		SA_{1-2}	Spindle running
Up		SA_{1-3}	Rocker arm rise
Down		SA_{1-4}	Rocker arm down

1) Zero voltage protection every time the power supply or power interruption after the resumption, must turn the cross switch to the left once. At this time, the microswitch contact SA_{1-1} is switched on, and the zero-voltage relay KA is absorbed and self-locked because the coil is energized. When the machine tool is working, the cross handle is not in the left position. At this time, if the power supply is cut off, the zero-voltage relay KA is released and its self-locking contact is also disconnected. When power is restored, relay KA not suck itself, control circuit not to electricity, this can prevent again after may occur when power is cut off, the machine to the risk of starting.

2) When the spindle motor is in operation, turn the cross switch to the right. The micro switch SA_{1-2} is closed. The contactless KM_1 is closed due to the current in the coil. The positive and negative rotation of the spindle is operated by a friction clutch handle on the headstock. The bit of a radial drilling machine rotates and moves up and down by a spindle motor. Switch the cross switch to the middle position, SA_{1-2} disconnect, spindle motor M_2 stop.

3) Rocker arm lifting bit and the relative height between the workpiece is not appropriate, rocker arm lifting can be adjusted. To make the rocker arm rise, the cross switch can be turned upwards. The micro-switch contact SA_{1-3} is closed. The contactor KM_2 pulls in because the coil is energized. The elastic mechanism of lifting screw and rocker arm is shown in Figure 4-5. When the lifting screw starts to turn forward, the rocker arm will not rise because the lifting nut also rotates. The lower auxiliary nut is moved upward because it cannot be rotated. The shaft of the transmission tightness device is rotated counterclockwise by the fork. As a result, the tightness device loosens the rocker arm. When the auxiliary nut moves up, the drive bar moves up. When the drive bar is pressed on the lifting nut, the lifting nut can no longer turn, but only drive the rocker arm up. When the auxiliary nut rises and the fork turns, the fork turns the shaft of the combination

switch SQ_2 to make the contact SQ_{2-2} close for clamping. At this point, the normally closed contact of KM_2 is disconnected, so the contactor KM_3 will not snap.

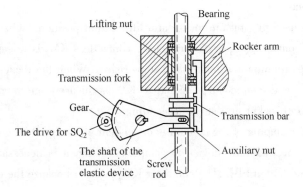

Figure 4-5 Schematic diagram of relaxation and tightening mechanism of rocker arm before and after lifting

When the rocker arm up to the required position, will cross switch back to the middle position, the contactor KM_2 released by coil power, the normally closed contact closure, and because of contact for SQ_{2-2} has been closed, contactor KM_3 off by coil electricity, motor M_3 to auxiliary nut moves down, on the one hand, drive transmission article moving up and down and nut out of contact, and lifting nut and screw idling, rocker arm stopped rising; On the other hand, when the auxiliary nut is lowered, the shaft of the transmission elastic device is turned clockwise by shifting fork. As a result, the elastic device clamps the rocker arm. At the same time, the fork rotates the shaft of the combination switch SQ_2 through the gear, so that the contact SQ_{2-2} is disconnected when the rocker arm is clamped, the contactor KM_3 is released, and the motor M_3 stops.

To lower the rocker arm, push the cross switch to the bottom and close the microswitch contact SA_{1-4}. The contactor KM_3 pulls in because the coil is powered on. The motor M_3 reverses and drives the lifting screw to reverse. At first, the lift nut also rotates so that the rocker arm does not descend. The lower auxiliary nut moves downward to turn the shaft of the transmission tightness device clockwise by shifting the fork. As a result, the tightness device also releases the rocker arm first. When the auxiliary nut moves downward, drive the drive bar to move downward. When the transmission bar pressure on the lifting nut, lifting nut also does not turn, drive the rocker arm down. When the auxiliary nut drops and rotates the fork, the fork turns the shaft of the combination switch SQ_2 to make the contact SQ_{2-1} close for clamping. In this case, the normally closed contact of KM_3 is disconnected.

When the position of the rocker arm down to need, will cross switch back to the middle position, then contact SA_{1-4} disconnect, contactor KM_3 released by coil power, the normally closed contact closure, and because of contact for SQ_{2-1} has been closed, contactor KM_2 for coil electricity and suck, motor M_3 moving turn auxiliary nut up, drive transmission article move up and down and nut out of contact, and lifting nut and screw idling, rocker arm stopped falling; When the auxiliary nut is moved up, the shaft of the transmission elastic device rotates counterclockwise through the shifting fork. As a result, the elastic device clamps the rocker arm. At the same time, the fork rotates the shaft of the combination switch SQ_2 through the gear, so that when the rocker arm is clamped, the contact SQ_{2-1} is disconnected, the contactor KM_2 is released, and the motor M_3 is stopped.

Limit switch SQ_1 is used to limit the rocker arm lifting limit position. When the rocker arm rises to the limit position, contact SQ_{1-1} is disconnected, contactless KM_2 is released due to coil power failure, motor M_3 stops running and rocker arm stops rising. When the rocker arm falls to the limit position, contact SQ_{1-2} is disconnected, contactless KM_3 is released due to coil power failure, motor M_3 stops running and rocker arm stops descending.

4) The release and clamping of the column is dependent on the positive and negative rotation of the motor M_4 through the hydraulic device to complete. When it is necessary to release the column, you can press the button SB_1. The contactor KM_4 pulls in because the coil is energized. The motor M_4 is in positive rotation. After release, the button SB_1 can be released. When the motor stops, the rocker arm and the outer column can be driven to rotate around the inner column by human power. When turning to the desired position, you can press SB_2, the contactor KM_5 pulls in due to the coil current, the motor M_4 reverses, and through the toothed clutch, M_4 drives the gear oil pump to reverse rotate, sending high pressure oil from the other direction, and clamping the column under the hydraulic push. After clamping, the button SB_2 can be released, the contactor KM_5 will be released due to the coil power failure, and the motor M_4 will stop.

The release and clamping of the spindle box of the Z35 radial drill press on the rocker arm and the release and clamping of the column are carried out by the same motor (M_4) and the same hydraulic mechanism.

(3) Analysis of Lighting Circuit the power supply of lighting circuit is provided by the transformer TC to reduce the AC voltage of 380V to the safe voltage of 36V. The lamp is grounded at the EL end for safety. The lamp is controlled by switch SA_3 and protected by fuse FU_3.

4.2.4.4 Failure Maintenance of Z35 Radial Drilling Machine

(1) Analysis and maintenance of common electrical faults.

1) Failure of spindle motor starting may be caused by:

① The FU_1 melt should be replaced if the melt is burnt out.

② Micro switch SA_{1-2} damage or bad contact, should be repaired or replaced.

③ The zero voltage relay KA has bad contact or loose connection, so the control circuit has no voltage.

④ The power supply voltage is too low.

⑤ The main contact of contactor KM_1 has bad contact or loose wiring.

2) The failure to stop the spindle motor is generally caused by the fusion welding of the normally open main contact of the contactor. The main contact of the contactor KM_1 should be replaced.

3) Failure of the rocker arm after lifting may be caused by:

① The position of SQ_2 moving contact of travel switch is offset. When the rocker arm has not been fully clamped after lifting and lifting, contact SQ_{2-1} (original rocker arm descending) or contact SQ_{2-2} (original rocker arm ascending) are disconnected prematurely, so it cannot be clamped completely. If the SQ_2 moving contacts SQ_{2-1} and SQ_{2-2} are adjusted to the appropriate positions, the fault can be eliminated.

② After the overhaul of the machine tool, the meshing position of the gear SQ_2 on the rotation stroke switch and the sector gear on the shifting fork was shifted. When the rocker arm was not fully clamped, the contact SQ_{2-1} or contact SQ_{2-2} were disconnected prematurely, and the motor M_3 stopped before the clamping position.

4) The lifting direction of the rocker arm is opposite to the lifting direction of the cross switch sign. The reason for this fault is that the power supply sequence of the lifting motor is reversed. This failure is dangerous and the power switch should be disconnected immediately. At this time, both the contact of the cross switch and the contact of the terminal limit switch are short-circuited by the contact of the travel switch SQ_2, losing control and terminal protection. To cross switch to the position of the rocker arm down as an example, the rocker arm lifting motor M_3 rocker arm after starting to rise without falling direction, then closed for SQ_{2-2}, cross switch back to zero, contact SQ_{1-4} disconnect, contactor KM_3 won't release, rocker arm, or continue to rise, until the limit switch will rise terminal knocked down, for SQ_{2-2} continues to open, rocker arm is still rising, shall be immediately disconnect the power supply.

5) Rocker arm lift cannot be stopped. Rocker arm lift to the required position, then the cross switch back to the middle position, rocker arm continues to lift, to the terminal limit switch contact disconnect also of no help. In this case, disconnect the power supply in time to avoid accidents. This is because the connection between SQ_{2-1} and SQ_{2-2} was accidentally switched during maintenance. Take the cross switch to the descending position as an example, KM_3 pulls in and the motor M_3 reverses, and the rocker arm first looses and then drops, after loosing, the contact SQ_{2-1} should be closed to prepare for clamping. When the connection is wrong, SQ_{2-2} will be closed, after that, the cross switch will be turned back to the middle position and the terminal limit switch contact SQ_{1-2} will be disconnected without stopping.

6) The rocker arm lifting motor operates alternately with positive and negative rotation. When the rocker arm lifting is completed, the rocker arm lifting motor M_3 shall reverse rotate to clamp the rocker arm. But if the trip switch for SQ_2 two contacts and for SQ_{2-1} for SQ_{2-2} too close, when the up (or down) to the desired location, cross switch back to zero, the contactor KM_2 (down for KM_3) has been released, contact for SQ_{2-2} (down for SQ_{2-1}) has been closed, KM_3 (reduced to KM_2), absorption motor reversal (reduced to forward) will radial clamping, clamping, for SQ_{2-2} (down for SQ_{2-1}) disconnect, KM_3 (reduced to KM_2) release. But because the mechanical inertia such as motor, motor and transmission part still move a short distance, make the trip switch contact for SQ_{2-1} (down for SQ_{2-2}) is connected by too close, contactor KM_2 (down for KM_3) and suction, motor and forward (down for inversion), and after a short motor M_3 for contact for SQ_{2-1} (down for SQ_{2-2}) disconnect and slow down, due to mechanical inertia round a short distance, make the trip switch contact for SQ_{2-2} (down for SQ_{2-1}) is connected by too close, contactor KM_3 (reduced to KM_2) off again, The motor reverses (down to positive) and then circulates,

making the clamping and loosening action repeated. The distance between the two contacts SQ_{2-1} and SQ_{2-2} of the travel switch should be carefully adjusted so that they are not too close and the fault can be eliminated.

7) Failure of column elastic motor can be caused by:

① Fuse FU_2 has broken, the melt should be replaced.

② Bad contact of button SB_1 or SB_2.

③ Bad contact of contactor KM_4 or KM_5.

8) The failure of the post tension motor to stop—generally due to the fusion welding of the main contact of the contortionists KM_4 or KM_5, the power supply should be disconnected immediately and the main contact should be replaced.

(2) Maintenance procedures and technological requirements.

1) Operate the drill press under the guidance of the operator, and understand various working conditions and operating methods of the drill press.

2) Under the guidance of teachers, make clear the installation location and wiring of drilling machine electrical components; Combine mechanical, electrical, hydraulic aspects of the relevant knowledge, understand the drill press electrical control special links.

3) Man-made installation of a natural fault on the Z35 radial drilling machine.

4) Teachers demonstrate maintenance. The steps are as follows:

① Guide the students to observe the failure phenomenon by using the electrified test method.

② According to the fault phenomenon, according to the circuit diagram with logical analysis to determine the fault range.

③ Use the correct inspection method, find the fault point and troubleshoot.

④ After maintenance, carry out the electrification test, and make maintenance records.

⑤ The teacher shall set up the fault points for the students to know in advance, guide the students how to analyze the fault phenomena, and gradually guide the students to adopt the correct maintenance steps and methods.

⑥ Teachers set up faults and students repair them.

(3) Matters needing attention.

1) Familiar with basic links and control requirements of Z35 radial drilling machine electrical circuit; Figure out how the electrical and actuator components work together to achieve a certain movement; Watch the teacher's demonstration overhaul carefully.

2) The tools and instruments used in maintenance shall meet the use requirements.

3) The original power phase sequence of the elevating motor cannot be changed at will.

4) When troubleshooting, the fault point must be repaired, but component substitution is not allowed.

5) During maintenance, it is strictly prohibited to extend the fault scope or generate new faults.

6) Live maintenance must be supervised by the instructor to ensure safety.

4.2.5 Bill of Materials

The table of the Bill of materials is shown in Table 4-6.

Table 4-6 Bill of materials

No.	Name	Number
1	XK-SX2C Advanced Maintenance Electrician Training Platform	1
2	Repair Electrician Training Components (XKDT11)	1
3	Repair Electrician Training Components (XKDT12A)	1
4	Three-phase Asynchronous Motor	1
5	Jumper	Several

4.2.6 Task Implementation

The table of the task implementation is shown in Table 4-7.

Table 4-7 Task implementation

Z35 radial drilling machine electrical wiring installation, commissioning-Task list

1. Complete the required device inventory and complete the following table as required

No.	Device	Number	Parameters	measurement	Remarks

2. Working principle description

3. Design and drawing of circuit wiring diagram

Continued Table 4-7

Z35 radial drilling machine electrical wiring installation, commissioning-Task list
4. Operation process record
5. Recording process of power-on test drive
6. Task summary

4.2.7 Task Evaluation

The table of the task evaluation is shown in Table 4-8.

Table 4-8 Task evaluation

Z35 radial drilling machine electric line installation, debugging rating table					
Device inventory and measurement (10 points)					
No.	Key inspection	Grading	Total score	Score	Remarks
1	Device inventory	1 point for one error	5		
2	Device measurement	1 point for one error	5		
		Subtotal			
Circuit diagram design (10 points)					
No.	Key inspection	Grading	Total score	Score	Remarks
1	Main circuit design	1 point for one error	5		
2	Control circuit design	1 point for one error	5		
		Subtotal			
Working principle description (10 points)					
No.	Key inspection	Grading	Total score	Score	Remarks
1	Working principle of main circuit	Complete discretion	5		
2	Working principle of control circuit	Complete discretion	5		
		Subtotal			

Continued Table 4-8

Z35 radial drilling machine electric line installation, debugging rating table

Line construction (30 points)

No.	Key inspection	Grading	Total score	Score	Remarks
1	Whether the wire is handled correctly	1 point for one error	10		
2	Whether the wire is firmly installed	1 point for one error	10		
3	whether the line correct	1 point for one error	10		
		Subtotal			

Power on test (30 points)

No.	Key inspection	Grading	Total score	Score	Remarks
1	Zero voltage protection function	Whether the function is realized	5		
2	Spindle control function	Whether the function is realized	5		
3	Rocker arm lifting control function	Whether the function is realized	10		
4	Column tension control function	Whether the function is realized	10		
		Subtotal			

Professional accomplishment (10 points)

No.	Key inspection	Grading	Total score	Score	Remarks
1	Power on operation	5 point for one error			
2	Standard operation	1 point for one error			
3	Team work	Discretion			
4	Work place tidy	Discretion			
		Total			

4.2.8 Troubleshooting

The table of the troubleshooting record is shown in Table 4-9.

Table 4-9 Troubleshooting record

	Troubleshooting record	
Fault 1	Fault phenomenon	
	Cause of fault	
	Trouble shooting process	

Continued Table 4-9

	Troubleshooting record	
Fault 2	Fault phenomenon	
	Cause of fault	
	Trouble shooting process	
Fault 3	Fault phenomenon	
	Cause of fault	
	Trouble shooting process	
Fault 4	Fault phenomenon	
	Cause of fault	
	Trouble shooting process	

4.2.9 Task Development

The faults of electrical equipment in the process of operation, some of which may be caused by human factors such as improper operation and use, unreasonable installation or incorrect maintenance, are called human faults. While some failure may be due to the electric equipment overload at run time, mechanical vibration, arc burning, long-term action of wear and tear, the influence of ambient temperature and humidity, and the erosion of harmful medium such as scrap metal and oil, quality problem or service life of electrical components and wait for a reason, referred to as the nature of failure. Obviously, if we strengthen the daily inspection, maintenance and maintenance of electrical equipment, timely find some abnormal factors, and give timely repair or replacement treatment, we can nip faults in the bud, nip them in the bud, make electrical equipment less or no faults, in order to ensure the normal operation of industrial machinery.

Routine maintenance of electrical equipment including motor and control equipment.

(1) Daily maintenance of the motor.

1) The surface of the motor shall be kept clean. Inlet and outlet shall be kept unimpeded. No water droplets, oil stains, metal scraps or any other foreign matter shall be allowed to fall into the motor.

2) Always check whether the load current of the motor in operation is normal. Use a clamp am-

meter to check whether the three-phase current is balanced. The difference between any phase of the three-phase current and its average value is not allowed to be more than 10%.

3) For motors operating under normal environmental conditions, the insulation resistance should be checked regularly with a megohm meter; For motors working in humid, dusty and containing corrosive gases and other environmental conditions, it is more appropriate to check their insulation resistance. Three-phase 380V motors and various low-voltage motors; The insulation resistance for at least 0.5MΩ before use. High voltage motor stator winding insulation resistance 1MΩ/kV; Rotor insulation resistance at least 0.5MΩ, just can use. If it is found that the insulation resistance of the motor does not meet the specified requirements, it shall take corresponding measures to make it meet the specified requirements before continuing to use.

4) Check the grounding device of the motor frequently to keep it firm and reliable.

5) Frequently check whether the supply voltage is consistent with the nameplate and whether the three-phase supply voltage is symmetric.

6) Regularly check whether the temperature rise of the motor is normal.

7) Regularly check whether the motor's vibration and noise are normal, whether there are abnormal smells, smoking, start-up difficulties and other phenomena. If found, stop immediately for maintenance.

8) Always check the motor bearing for overheating, insufficient grease or wear, and the vibration and axial displacement of the bearing shall not exceed the specified value. The bearing should be cleaned and inspected regularly, and the bearing grease should be replenished or replaced regularly (generally about one year).

9) For the wound rotor asynchronous motor, check the contact pressure, wear and spark between the brush and the slip ring. When abnormal sparks are found, further check the brush or clean the surface of the slip ring, and correct the brush spring pressure. There should be 2~4mm spacing between brush grip and slip ring; Brush and brush grip inner wall should keep 0.1~0.2mm clearance; For severe wear and tear should be replaced.

10) Check whether the commutator surface of dc motor is smooth and round, and whether there is mechanical damage or spark burn. If touch with the sundry such as carbon powder, grease, want to dip in alcohol with clean soft cotton cloth to wipe. After the commutator operates for a long time under load, a layer of uniform dark brown oxide film will be generated on its surface, which has the function of protecting the commutator. Abrasive cloth should be avoided by all means. But when the commutator surface appear obvious burning marks or spark burning appear uneven phenomenon, may need to be done on its surface with zero emery cloth careful grinding or with lathe car light again, then the commutator slice of mica under carved between 1~1.5mm deep, and will be on the surface of the burr, after cleaning to clean, used to reassemble.

11) Check whether the mechanical transmission device is normal and whether the coupling, belt wheel or transmission gear is runout.

12) Check whether the outgoing wires of the motor are well insulated and connected reliably.

(2) Control the daily maintenance of equipment.

1) The door, cover, lock and the oil-resistant gasket around the door frame of the electric cabi-

net shall be in good condition. The door, cover should be closed tightly, the cabinet should be kept clean, no water, oil and metal debris into the cabinet, so as not to damage the electrical appliances cause accidents.

2) All control buttons, handle of main command switch, signal lamp and instrument cover on the control table shall be kept clean and intact.

3) Check whether the contact system of contactors, relays and other electrical appliances has good suction, whether there is noise, jamming or hysteresis, and whether the contact surface has ablation, burr or holes; Whether the electromagnetic coil overheats; Whether all kinds of spring force is appropriate; Whether the arcing extinguisher is intact or not.

4) Test whether the position switch can play a role of position protection.

5) Check whether the operating mechanism of each electrical appliance is flexible and reliable, and whether the setting value meets the requirements.

6) Check whether the connection between each line joint and terminal plate is solid, the connecting wires, cables or hoses protecting wires between each component shall not be corroded by coolant, oil, etc. , and the pipe joint shall not fall off or fall apart.

7) Check whether the heat dissipation condition of the electric cabinet and conductor channel is in good condition.

8) Check whether all kinds of indicating signal devices and lighting devices are in good condition.

9) Check whether all bare conductors on electrical equipment and industrial machinery are connected to the special terminals for protection grounding.

(3) Maintenance cycle of electrical equipment.

Setin the electrical components in the electrical cabinet, generally do not often open the door monitoring, mainly by regular maintenance, to achieve a longer period of safe and stable operation of electrical equipment. The maintenance cycle shall be determined according to the structure, use and environmental conditions of the electrical equipment. Generally, the maintenance of electrical equipment can be carried out at the same time with the first and second level maintenance of industrial machinery.

4.2.10 Task Summary

Taking the Z35 radial drilling machine electrical line installation, debugging and fault maintenance training as an example, further understand the electrical fault maintenance and repair matters needing attention of the machine tool, and analyze and find out the root cause of the fault in the maintenance; Avoid extended failure; After finding out fault point, must aim at different fault circumstance and position, adopt correct repair method, do not easily adopt the method such as replacing electric element and filling line, more do not allow to change circuit easily or replace electric element with different specification, in order to prevent generation of human fault. When the electrical fault is repaired and the power is needed for trial operation, it shall cooperate with the operator. After each fault is removed, experience shall be summarized in time and maintenance records shall be kept.

References

[1] Cao Shiyong. Research on Fault Inspection and Maintenance Methods of Low Voltage Electrical Appliances [J]. Science and Technology Wind, 2019.

[2] Wang Lianyi. Discussion on The Implementation of Condition Monitoring and Maintenance Mode for Low-Voltage Electrical Equipment [J]. Internal Combustion Engine and Accessories, 2018.

[3] Lv Yuming. Maintenance Electrician Skill Training [M]. Beijing: Beijing Normal University Press, 2016.

[4] Wang Bing. National Professional Appraisal Guide for Maintenance Electrician [M]. Beijing: Electronic Industry Press, 2012.

[5] Yang Zongqiang. One Book on Electrical Maintenance Skills [M]. Beijing: Chemical Industry Press, 2019.

[6] Han Xuetao. Full Coverage of Electrical Maintenance [M]. Beijing: Electronic Industry Press, 2019.

[7] Cao Jinhong. New Practical Electrician Manual [M]. Tianjin: Tianjin Science and Technology Press, 2018.

[8] Ren Qingchen. Design and Production of Electrical Control Cabinet [M]. Beijing: Electronic Industry Press, 2014.

[9] Bai Gong. Manual of Maintenance Electrician Skills [M]. Beijing: China Machine Press, 2017.

[10] Zhu Zhaohong. Basic Skills of Maintenance Electrician [M]. Beijing: China Labor and Social Security Press, 2019.

[11] Li Shuhai. Electrical Low Voltage Operation and Maintenance [M]. Beijing: Chemical Industry Press, 2011.

[12] Qin Zhongquan. Low Voltage Electrician Skill Book [M]. Beijing: Chemical Industry Press, 2012.

[13] Sun Yang. Selected 300 Practical Electrical Circuits [M]. Beijing: Chemical Industry Press, 2019.

[14] Qin Zhongquan. Low Voltage Electrical Practical Skills Book [M]. Beijing: Chemical Industry Press, 2017.

[15] Wang Jianhua. Handbook of Electrical Engineers [M]. Beijing: China Machine Press, 2019.

[16] Zhang Baifan. Principle and Control Technology of Low Voltage Switchgear [M]. Beijing: China Machine Press, 2019.

[17] Zhang Xiaojiang. Fundamentals of Electrical Machinery and Drive [M]. Beijing: China Machine Press, 2016.

[18] Tang Jie. Motor and Drag [M]. Beijing: Higher Education Press, 2020.

[19] Chen Baoling. Practical Training Course of Electric Motor and Electric Control [M]. Beijing: Beijing Normal University Press, 2014.

[20] Cheng Zhou. Motor Drive and Electronic Control Technology [M]. Beijing: Electronic Industry Press, 2013.

项目1 电工基本技能训练

任务1.1 电工安全技术与触电急救

1.1.1 任务描述

电在造福于人类的同时,也会给人类带来灾难。在电气工伤事故中,触电事故占的比例相当大。因此学习必要的安全用电知识以及触电急救措施是十分必要的。

1.1.2 任务目标

(1) 认识人体触电的基本知识。
(2) 掌握电工安全操作及安全用电。
(3) 了解触电急救的常识及现场急救方法。

1.1.3 任务分析

本任务分为:
(1) 触电的基本知识;
(2) 电工安全操作规程及安全用电常识;
(3) 触电急救的常识及现场急救方法。

1.1.4 任务相关知识

1.1.4.1 触电的基本知识

人体与电接触时,对人体各部位组织(如皮肤、心脏、呼吸器官和神经系统)不会造成任何损害的电压叫作安全电压。

安全电压值的规定,各国有所不同。如荷兰和瑞典为24V,美国为40V,法国交流为24V、直流为50V,波兰为50V。

(1) 跨步电压触电。当带电设备发生某相接地时,接地电流流入大地,在距接地点不同的地表面各点上呈现不同电位,电位的高低与离开接地点距离有关,距离越远,电位越低。

当人的脚与脚之间同时踩在带有不同电位的地表面两点时,会引起跨步电压触电。如果遇到这种危险场合,应合拢双脚跳离接地处20m之外,以保障人身安全。

(2) 相间触电。所谓相间触电就是在人体与大地绝缘的时候,同时接触两根不同的相线或人体同时接触电气设备不同相的两个带电部分时,这时电流由一根相线经过人体到另一个相线,形成闭合回路。这种情形称为相间触电,此时人体直接处在线电压作用之下,比单相触电的危险性更大。

人体触电示意图如图1-1所示。

图1-1 人体触电示意图

（3）致命电流。在较短的时间内危及生命的最小电流称为致命电流，在电流不超过百毫安的情况下，电击致命的主要原因是电流引起心室颤动或窒息造成的。因此，可以认为引起心室颤动的电流即为致命电流。

（4）人体触电时的危险性与相关因素。

1）人体触电时，致命的因素是通过人体的电流，而不是电压，但是当电阻不变时，电压越高，通过导体的电流就越大。因此，人体触及带电体的电压越高，危险性越大。但不论是高压还是低压，触电都是危险的。

2）电流通过人体的持续时间是影响电击伤害程度的又一重要因素。人体通过电流的时间越长，人体电阻就越降低，流入的电流就越大，后果就越严重。另一方面，人的心脏每收缩、扩张一次，中间约有0.1s间歇，这0.1s对电流最敏感。如果电流在这一瞬间通过心脏，即使电流很小（零点几毫安），也会引起心脏震颤；如果电流不在这一瞬间通过，即使电流较大，也不至于引起心脏停搏。由此可知，如果电流持续时间超过0.1s，则必然与心脏最敏感的间隙相重合而造成很大的危险。

3）电流通过人体的途径也与电击伤程度有直接关系。电流通过人体头部，会使人立即昏迷，电流如果通过脊髓会使人半截肢体瘫痪，电流通过心脏、呼吸系统和中枢神经，会引起精神失常或引起心脏停止跳动，中断全身血液循环，造成死亡。因此，从手到脚的电流途径最为危险。其次，是手到手的电流途径，再次是脚到脚的电流途径。

4）电流频率对电击伤害程度有很大影响。50Hz的工频交流电对设计电气设备比较合理，但是这种频率的电流对人体触电伤害程度也最严重。

5）人的健康状况，人体皮肤的干湿等情况对电击伤害程度也有一定影响。凡患有心脏病，神经系统疾病或结核病的病人电击伤害程度比健康人严重。此外，皮肤干燥电阻大,通过的电流小,皮肤潮湿电阻小,通过的电流就大,危害也大。

(5) 接地保护。接地保护又常称为保护接地,就是将电气设备的金属外壳与接地体连接,以防止因电气设备绝缘损坏而使外壳带电时,操作人员接触设备外壳而触电。

(6) 接触电势、接触电压、跨步电势和跨步电压。当接地短路电流流过接地装置时,大地表面形成分布电位,在地表面上离设备水平距离为 0.8m 处与沿设备外壳、构架或墙壁垂直距离 1.8m 处两点间的电位差称为接触电势。人体接触该两点时所承受的电压称为接触电压;接地网网孔中心对接地网接地体的最大电位差称为最大接触电势,人体接触该两点时所承受的电压称为最大接触电压。地面上水平距离为 0.8m 的两点间的电位差称为跨步电势。人体两脚接触该两点时所承受的电压称为跨步电压;接地网外的地面上水平距离 0.8m 处对接地网边缘接地体的电位差称为最大跨步电势,人体两脚接触该两点时所承受的电压称为最大跨步电压。

发生触电时,流经人体的电流决定于触电电压与人体电阻的比值。人体电阻并不是一个固定数值。人体各部分的电阻除去角质层外,以皮肤的电阻最大。当人体在皮肤干燥和无损伤的情况下,人体的电阻可高达 $40000 \sim 400000 \Omega$。如果除去皮肤,则人体电阻可下降至 $600 \sim 800 \Omega$。但人体的皮肤电阻也并不是固定不变的,当皮肤出汗潮湿或是受到损伤时,电阻就会下降到 1000Ω 左右。

感知电流:用手握住电源时,手心感觉发热的直流电流,或因神经受刺激而感觉轻微刺痛的交流电流称为感知电流。受试者双手放在小铜丝上面,直流电流的平均感知电流男性是 5.2mA,女性是 3.5mA。

摆脱电流:触电后能自行摆脱的电流称为摆脱电流。由测定结果得知,男性的工频摆脱电流是 9mA,女性是 6mA。当 $18 \sim 22mA$(摆脱电流的上限)的工频电流通过人体的胸部时,所引起的肌肉反应将使触电者在通电时间内停止电流,呼吸即可恢复,而且不会因短暂的呼吸停止而造成不良后果。

1.1.4.2 电工安全操作规程及安全用电常识

(1) 电工安全操作知识。
1) 用电仪器设备的金属外壳应有良好的接地线。
2) 按规定使用和更换熔断丝,不可以随意加大规格,以免烧毁所用仪器设备。
3) 操作现场(地)要铺设胶垫。
4) 不违章操作,养成单手操作的良好习惯。
5) 有特殊情况首先拉闸断电。
6) 经常对设备进行安全检查,检查有无裸露的带电部分和漏电情况。裸露的带电线头,必须及时地用绝缘材料包好。检验时,应使用专用的验电设备,任何情况下都不要用手去鉴别。
7) 装设保护接地或保护接零。当设备的绝缘损坏,电压窜到其金属外壳时,把外壳上的电压限制在安全范围内,或自动切断绝缘损坏的电气设备。
8) 正确使用各种安全用具,如绝缘棒、绝缘夹钳、绝缘手套、绝缘套鞋、绝缘地毯等。并悬挂各种警告牌,装设必要的信号装置。
9) 安装漏电自动开关。当设备漏电、短路、过载或人身触电时,自动切断电源,对设备和人身起保护作用。

10）当停电检修时及接通电源前都应采取措施使其他有关人员知道，以免有人正在检修时，其他人合上电闸；或者在接通电源时，其他人员由于不知道而正在作业，造成触电。

11）中性点直接接地的低压三相380V电网的用电设备外壳采用保护接零，禁止采用保护接地。

（2）电气安全技术措施。在电气设备和线路上工作，尤其是在高压场所工作，必须完成停电、验电、放电、装设临时接地线、悬挂警告牌、装设遮栏等保证安全的技术措施。

1）停电。对所有可能来电的线路，要全部切断，且应有明显的断开点。要特别注意防止从低压侧向被检修设备反送电，要采取防止误合闸的措施。

2）验电。对已停电的线路要用与电压等级相适应的验电器进行验电。

3）放电。其目的是消除被检修设备上残存的电荷。放电可用绝缘棒或开关来进行操作。应注意线与地之间，线与线之间均应放电。

4）装设临时接地线。为防止作业过程中意外送电和感应电，要在检修的设备和线路上装设临时接地线和短路线。

5）悬挂警告牌和装设遮栏。在被检修的设备和线路的电源开关上，应加锁并悬挂"有人作业，禁止送电"的警告牌。对于部分停电的作业，安全距离小于0.7m的未停电设备，应装设临时遮栏并悬挂"止步，高压危险"的标示牌等。

电气安全技术措施如图1-2所示。

图1-2 电气安全技术措施

1.1.4.3 触电急救的常识及现场急救方法

触电者脱离电源后，应立即就地进行抢救。"立即"之意就是争分夺秒，不可贻误。"就地"之意就是不能消极地等待医生的到来，而应在现场施行正确救护的同时，派人通知医务人员到现场并做好将触电者送往医院的准备工作。

触电急救措施如图1-3所示。

（1）现场抢救措施。

1）触电者未失去知觉的救护措施。如果触电者所受的伤害不太严重，神志尚清醒，

图 1-3 触电急救措施

只是心悸、头晕、出冷汗、恶心、呕吐、四肢发麻、全身乏力,甚至一度昏迷,但未失去知觉,则应让触电者在通风暖和的处所静卧休息,并派人严密观察,同时请医生前来或送往医院诊治。

2)触电者已失去知觉(心肺正常)的抢救措施。如果触电者已失去知觉,但呼吸和心跳尚正常,则应使其舒适地平卧着,解开衣服以利呼吸,四周不要围人,保持空气流通,冷天应注意保暖,同时立即请医生前来或送往医院诊察。若发现触电者呼吸困难或心跳失常,应立即施行人工呼吸或胸外心脏按压。

3)对"假死"者的急救措施。如果触电者呈现"假死"(即所谓电休克)现象,则可能有3种临床症状:一是心跳停止,但尚能呼吸;二是呼吸停止,但心跳尚存(脉搏很弱);三是呼吸和心跳均已停止。"假死"症状的判定方法是"看""听""试"。"看"是观察触电者的胸部、腹部有无起伏动作;"听"是用耳贴近触电者的口鼻处,听他有无呼气声音;"试"是用手或小纸条试测口鼻有无呼吸的气流,再用两手指轻压一侧(左或右)喉结旁凹陷处的颈动脉有无搏动感觉。如"看""听""试"的结果,既无呼吸又无颈动脉搏动,则可判定触电者呼吸停止或心跳停止或呼吸心跳均停止。

(2)心肺复苏。当判定触电者呼吸和心跳停止时,应立即按心肺复苏法就地抢救。所谓心肺复苏法就是支持生命的三项基本措施,即通畅气道、口对口(鼻)人工呼吸、胸外按压(人工循环)。

1)通畅气道。

①清除口中异物。使触电者仰面躺在平硬的地方,迅速解开其领扣、围巾、紧身衣和裤带。如发现触电者口内有食物、假牙、血块等异物,可将其身体及头部同时侧转,迅速用一个手指或两个手指交叉从口角处插入,从中取出异物,操作中要注意防止将异物推到咽喉深处。

②采用仰头抬颌法,通畅气道。操作时,救护人用一只手放在触电者前额,另一只手的手指将其颏颌骨向上抬起,两手协同将头部推向后仰,舌根自然随之抬起、气道即可畅通。为使触电者头部后仰,可于其颈部下方垫适量厚度的物品,但严禁用枕头或其他物品垫在触电者头下,因为头部抬高前倾会阻塞气道,还会使施行胸外按压时流向脑部的血量减小,甚至完全消失。

仰头抬颌法示意图如图 1-4 所示。

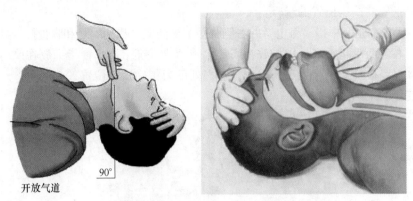

图 1-4　仰头抬颌法示意图

2) 口对口（鼻）人工呼吸。救护人在完成气道通畅的操作后，应立即对触电者施行口对口或口对鼻人工呼吸。口对鼻人工呼吸用于触电者嘴巴紧闭的情况。

先大口吹气刺激起搏，救护人蹲跪在触电者的左侧或右侧；用放在触电者额上那只手的手指捏住其鼻翼，另一只手的食指和中指轻轻托住其下巴；救护人深吸气后，与触电者口对口紧合，在不漏气的情况下，先连续大口吹气两次，每次 1~1.5s；然后用手指试测触电者颈动脉是否有搏动，如仍无搏动，可判断心跳确已停止，在施行人工呼吸的同时应进行胸外按压。

大口吹气两次试测颈动脉搏动后，立即转入正常的口对口人工呼吸阶段。正常的吹气频率是每分钟约 12 次。正常的口对口人工呼吸操作姿势如上述。但吹气量不需过大，以免引起胃膨胀，如触电者是儿童，吹气量宜小些，以免肺泡破裂。救护人换气时，应将触电者的鼻或口放松，让他借自己胸部的弹性自动吐气。吹气和放松时要注意触电者胸部有无起伏的呼吸动作。吹气时如有较大的阻力，可能是头部后仰不够，应及时纠正，使气道保持畅通。触电者如牙关紧闭，可改行口对鼻人工呼吸。吹气时要将触电者嘴唇紧闭，防止漏气。

口对口人工呼吸示意图如图 1-5 所示。

图 1-5　口对口人工呼吸示意图

3) 胸外按压。胸外按压是借助人力使触电者恢复心脏跳动的急救方法。其有效性在于选择正确的按压位置和采取正确的按压姿势。

①确定正确的按压位置的步骤：

第一，手的食指和中指沿触电者的右侧肋弓下缘向上，找到肋骨和胸骨接合处的中点。

第二，右手两手指并齐，中指放在切迹中点（剑突底部），食指平放在胸骨下部，另一只手的掌根紧挨食指上缘置于胸骨上，掌根处即为正确按压位置。

胸外按压的正确位置如图 1-6 所示。

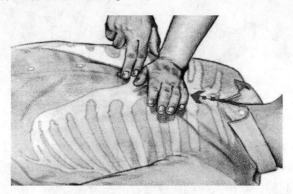

图 1-6　胸外按压的正确位置

②正确的按压姿势：

第一，使触电者仰面躺在平硬的地方并解开其衣服，仰卧姿势与口对口（鼻）人工呼吸法相同。

第二，救护人立或跪在触电者一侧肩旁，两肩位于触电者胸骨正上方，两臂伸直，肘关节固定不屈，两手掌相叠，手指翘起，不接触触电者胸壁。

第三，以髋关节为支点，利用上身的重力，垂直将正常成人胸骨压陷 3～5cm（儿童和瘦弱者酌减）。

第四，至要求程度后，立即全部放松，但救护人的掌根不得离开触电者的胸壁。按压有效的标志是在按压过程中可以触到颈动脉搏动。

胸外按压的正确姿势如图 1-7 所示。

③恰当的按压频率：

第一，胸外按压要以均匀速度进行。操作频率以每分钟 80 次为宜，每次包括按压和放松一个循环，按压和放松的时间相等。

第二，当胸外按压与口对口（鼻）人工呼吸同时进行时，操作的节奏为：单人救护时，每按压 15 次后吹气 2 次（15∶2），反复进行；双人救护时，每按压 15 次后由另一人吹气 1 次（15∶1），反复进行。

（3）现场救护中的注意事项。

1）抢救过程中应适时对触电者进行再判定。按压吹气 1 分钟后（相当于单人抢救时做了 4 个 15∶2 循环），应采用"看、听、试"方法在 5～7s 钟内完成对触电者是否恢复自然呼吸和心跳的再判断。若判定触电者已有颈动脉搏动，但仍无呼吸，则可暂停胸外按压，而再进行两次口对口人工呼吸，接着每隔 5s 吹气一次（相当于每分钟 12 次）。如果脉搏和呼吸仍未能恢复，则继续坚持心肺复苏法抢救。在抢救过程中，要每隔数分钟用

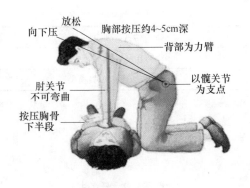

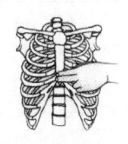 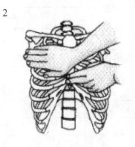 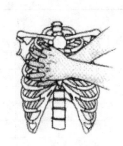

图1-7 胸外按压的正确姿势

"看、听、试"方法再判定一次触电者的呼吸和脉搏情况,每次判定时间不得超过5~7s。在医务人员未前来接替抢救前,现场人员不得放弃现场抢救。

2) 抢救过程中移送触电伤员时的注意事项:

①心肺复苏应在现场就地坚持进行,不要图方便而随意移动触电伤员,如确有需要移动时,抢救中断时间不应超过30s。

②移动触电者或将其送往医院,应使用担架并在其背部垫以木板,不可让触电者身体蜷曲着进行搬运。移送途中应继续抢救,在医务人员未接替救治前不可中断抢救。

③应创造条件,用装有冰屑的塑料袋作成帽状包绕在伤员头部,露出眼睛,使脑部温度降低,争取触电者心、肺、脑能得以复苏。

④如触电者的心跳和呼吸经抢救后均已恢复,可暂停心肺复苏法操作。但心跳呼吸恢复的早期仍有可能再次骤停,救护人应严密监护,不可麻痹,要随时准备再次抢救。触电者恢复之初,往往神志不清、精神恍惚或情绪躁动、不安,应设法使他安静下来。

1.1.5 材料清单

材料清单见表1-1。

表1-1 材料清单

序号	名 称	数量
1	XK-SX2C型高级维修电工实训台	1台
2	维修电工实训组件(XKDT11)	1个
3	维修电工实训组件(XKDT12A)	1个

续表 1-1

序号	名　称	数量
4	防止触电警示牌	1块
5	胸外按压模拟示教模型	1个

1.1.6 任务实施

任务书见表 1-2。

表 1-2　任务书

(　　　　　　　　　　　　　　　　) 任务书				

1. 根据要求完成所需器件清点并完成下表

序号	器件名称	器件数量	器件参数	器件测量	备注

2. 工作任务描述

3. 操作过程记录

续表 1-2

( )任务书
4. 任务小结

1.1.7 任务评价

评分表见表 1-3。

表 1-3 评分表

( )评分表					
器件清点和测量（10 分）					
序号	重点检查内容	评分标准	分值	得分	备注
1	器件清点	器件清点错误一项扣 1 分	5		
2	器件摆放	器件摆放错误一项扣 1 分	5		
小　计					
安全用电警示牌正确使用（10 分）					
序号	重点检查内容	评分标准	分值	得分	备注
1	停送电安全操作规范	错误一处扣 1 分	5		
2	安全警示牌用法	错误一处扣 1 分	5		
小　计					
口对口人工呼吸正确操作（25 分）					
序号	重点检查内容	评分标准	分值	得分	备注
1	教具姿势摆放	姿势是否正规	5		
2	操作步骤	步骤是否完整	20		
小　计					
胸外按压的正确操作（25 分）					
序号	重点检查内容	评分标准	分值	得分	备注
1	教具姿势摆放	错误一处扣 1 分	5		
2	按压位置	错误一处扣 1 分	10		
3	按压姿势	错误一处扣 1 分	10		
小　计					

续表 1-3

（　　　　　　　　　　　　　　　）评分表

现场触电急救处理（20分）

序号	重点检查内容	评分标准	分值	得分	备注
1	对触电者的判定	是否了解	5		
2	移送伤员	是否了解	5		
3	对伤员恢复后的处理	是否了解	10		
	小　　计				

职业素养（10分）

序号	重点检查内容	评分标准	扣分	得分	备注
1	用电安全意识	一次扣5分扣完为止			
2	规范操作	一次扣1分扣完为止			
3	团队合作	酌情			
4	工位整洁	酌情			
	总　　计				

1.1.8　任务拓展

电气火灾是指由电气原因引发燃烧而造成的灾害。短路、过载、漏电等电气事故都有可能导致火灾。设备自身缺陷、施工安装不当、电气接触不良、雷击静电引起的高温、电弧和电火花是导致电气火灾的直接原因。周围存放易燃易爆物是电气火灾的环境条件。

电气火灾产生的直接原因：

（1）设备或线路发生短路故障。电气设备由于绝缘损坏、电路年久失修、疏忽大意、操作失误及设备安装不合格等将造成短路故障，其短路电流可达正常电流的几十倍甚至上百倍，产生的热量（正比于电流的平方）使温度上升超过自身和周围可燃物的燃点引起燃烧，从而导致火灾。

（2）过载引起电气设备过热。选用线路或设备不合理，线路的负载电流量超过了导线额定的安全载流量，电气设备长期超载（超过额定负载能力），引起线路或设备过热而导致火灾。

（3）接触不良引起过热。如接头连接不牢或不紧密、动触点压力过小等使接触电阻过大，在接触部位发生过热而引起火灾。

（4）通风散热不良。大功率设备缺少通风散热设施或通风散热设施损坏造成过热而引发火灾。

（5）电器使用不当。如电炉、电熨斗、电烙铁等未按要求使用，或用后忘记断开电源，引起过热而导致火灾。

（6）电火花和电弧。有些电气设备正常运行时就能产生电火花、电弧，如大容量开关、接触器触点的分、合操作，都会产生电弧和电火花。电火花温度可达数千度，遇可燃物便可点燃，遇可燃气体便会发生爆炸。

日常生活和生产的各个场所中，广泛存在着易燃易爆物质，如石油液化气、煤气、天然气、汽油、柴油、酒精、棉、麻、化纤织物、木材、塑料等，另外一些设备本身可能会产生易燃易爆物质，如设备的绝缘油在电弧作用下分解和气化，喷出大量油雾和可燃气体；酸性电池排出氢气并形成爆炸性混合物等。一旦这些易燃易爆环境遇到电气设备和线路故障导致的火源，便会立刻着火燃烧。

电气火灾的防护措施主要致力于消除隐患、提高用电安全，具体措施如下：

（1）正确选用保护装置，防止电气火灾发生。对正常运行条件下可能产生电热效应的设备采用隔热、散热、强迫冷却等结构，并注重耐热、防火材料的使用。按规定要求设置包括短路、过载、漏电保护设备的自动断电保护。对电气设备和线路正确设置接地、接零保护，为防雷电安装避雷器及接地装置。根据使用环境和条件正确设计选择电气设备。恶劣的自然环境和有导电尘埃的地方应选择有抗绝缘老化功能的产品，或增加相应的措施；对易燃易爆场所则必须使用防爆电气产品。

（2）正确安装电气设备，防止电气火灾发生。合理选择安装位置。对于爆炸危险场所，应该考虑把电气设备安装在爆炸危险场所以外或爆炸危险性较小的部位。开关、插座、熔断器、电热器具、电焊设备和电动机等应根据需要，尽量避开易燃物或易燃建筑构件。起重机滑触线下方不应堆放易燃品。露天变、配电装置不应设置在易于沉积可燃性粉尘或纤维的地方等。

保持必要的防火距离。对于在正常工作时能够产生电弧或电火花的电气设备，应使用灭弧材料将其全部隔围起来，或将其与可能被引燃的物料，用耐弧材料隔开或与可能引起火灾的物料之间保持足够的距离，以便安全灭弧。

安装和使用有局部热聚焦或热集中的电气设备时，在局部热聚焦或热集中的方向与易燃物料，必须保持足够的距离，以防引燃。

电气设备周围的防护屏障材料必须能承受电气设备产生的高温（包括故障情况下）。应根据具体情况选择不可燃、阻燃材料或在可燃性材料表面喷涂防火涂料。

保持电气设备的正常运行，防止电气火灾发生。正确使用电气设备是保证电气设备正常运行的前提。因此应按设备使用说明书的规定操作电气设备。严格执行操作规程。保持电气设备的电压、电流、温升等不超过允许值。保持各导电部分连接可靠，接地良好。保持电气设备的绝缘良好，保持电气设备的清洁，保持良好通风。

1.1.9 任务小结

安全用电包括供电系统的安全、用电设备的安全及人身安全3个方面，它们之间又是紧密联系的。供电系统的故障可能导致用电设备的损坏或人身伤亡事故，而用电事故也可能导致局部或大范围停电，甚至造成严重的社会灾难。

在用电过程中，必须特别注意电气安全，如果稍有麻痹或疏忽，就可能造成严重的人身触电事故，或者引起火灾或爆炸，给国家和人民带来极大的损失。

任务1.2 数字万用表的使用

1.2.1 任务描述

在电工专业的各种工作环境中，经常需要工作人员对设备进行必要的监测和测量，这就需要我们使用到各种测量仪器。数字万用表相对来说属于比较简单的测量仪器。它所具有的特点包括：输入阻抗高，输入电容小，因而对被测电路的影响较小；测量频率范围宽，一般可从几十赫至几兆赫；灵敏度高；电压测量范围较大，低到微伏级或毫伏级，高则可达上千伏和几千伏；指示直观，数字万用表在测量时可由液晶显示屏直接读出电量数值，既方便又直观；功能齐全，具有普通型万用表的全部功能以及其他附加功能（如测量晶体管等有关参数）。

本次任务重点教大家数字万用表的正确使用方法。从数字万用表的电压、电阻、电流、二极管、晶体管、MOS 场效应晶体管的测量等测量方法开始，让大家更好地掌握万用表测量方法。

1.2.2 任务目标

（1）掌握数字万用表的使用方法。
（2）了解数字万用表在使用过程中的注意事项。

1.2.3 任务分析

数字万用表是一种多用途电子测量仪器，一般包含安培计、电压表、欧姆计等功能，有时也称为万用计、多用计、多用电表或三用电表。数字万用表有用于基本故障诊断的便携式装置，也有放置在工作台的装置，有的分辨率可以达到七、八位。数字多用表（DMM）就是在电气测量中要用到的电子仪器。它可以有很多特殊功能，但主要功能就是对电压、电阻和电流进行测量，数字多用表，作为现代化的多用途电子测量仪器，主要用于物理、电气、电子等测量领域。

本任务分为：
（1）电压的测量；
（2）电流的测量；
（3）电阻的测量；
（4）二极管的测量；
（5）蜂鸣通断测试；
（6）晶体管的测量；
（7）MOS 场效应晶体管的测量；
（8）使用数字万用表的注意事项。

1.2.4 任务相关知识

数字万用表的工作原理是将被测量信号经过放大之后，再经过数字化处理，最终将测量结果由数字表头以数字的形式显示出来的一种万用表，它的测量准确度高、分辨力高、电压灵敏度高、测量种类多、功能齐全、过载能力强、抗干扰性好、体积小、重量轻、可靠性高，又由于采用数字形式来显示测量结果，使得读数快捷方便，而且还能从根本上消除因视差所造成的读数误差，因此在电气、电子、通信、科研和家用电器行业等的应用非常广泛。

数字万用表外观功能示意图如图 1-8 所示。
数字万用表符号说明如图 1-9 所示。

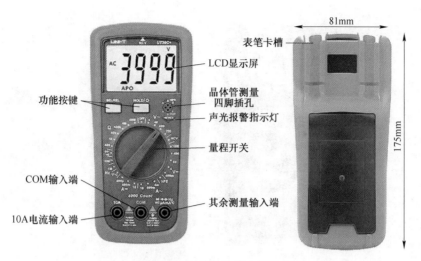

图 1-8 数字万用表外观功能示意图

符号	功能
V∼	交流电压测量
V⎓	直流电压测量
A∼	交流电流测量
A⎓	直流电流测量
Ω	电阻测量
Hz	频率测量
h_{FE}	晶体管测量
F	电容测量
℃	温度测量
⇥	二极管测量
·)))	通断测量

图 1-9 数字万用表符号说明

1.2.4.1 电压的测量

（1）直流电压的测量。如电池、随身听电源等。首先将黑表笔插进"com"孔，红表笔插进"V Ω"。把旋钮选到比估计值大的量程（注意：表盘上的数值均为最大量程，"V-"表示直流电压档，"V~"表示交流电压档，"A"是电流档），接着把表笔接电源或电池两端；保持接触稳定。数值可以直接从显示屏上读取，若显示为"1."，则表明量程太小，那么就要加大量程后再测量工业电器。如果在数值左边出现"-"，则表明表笔极性与实际电源极性相反，此时红表笔接的是负极。

数字万用表测直流电压如图 1-10 所示。

（2）交流电压的测量。表笔插孔与直流电压的测量一样，不过应该将旋钮打到交流档"V~"处所需的量程即可。交流电压无正负之分，测量方法跟前面相同。无论测交流还是

图 1-10 数字万用表测直流电压

直流电压,都要注意人身安全,不要随便用手触摸表笔的金属部分。

数字万用表测交流电压如图 1-11 所示。

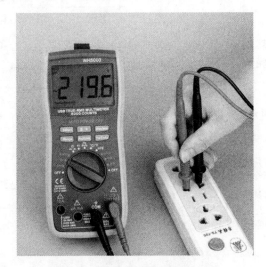

图 1-11 数字万用表测交流电压

1.2.4.2 电流的测量

(1) 直流电流的测量。先将黑表笔插入"COM"孔。若测量大于 200mA 的电流,则要将红表笔插入"20A"插孔并将旋钮打到直流"20A"档;若测量小于 200mA 的电流,则将红表笔插入"mA"插孔,将旋钮打到直流 200mA 以内的合适量程。调整好后,就可以测量了。将万用表串进电路中,保持稳定,即可读数。若显示为"1",那么就要加大量程;如果在数值左边出现"-",则表明电流从黑表笔流进万用表。插孔"A"有 200mA 熔断器保护,过载会将熔体熔断,应按规定值及时更换。插孔"20A"无熔断器保护,可连续测量的最大电流为 10A,测量时间应小于 15s。

（2）交流电流的测量。测量方法与直流电流的测量方法相同，不过档位应该打到交流档位。电流测量完毕后应将红笔插回"VΩ"孔，若忘记这一步而直接测电压，会损坏万用表及设备。

数字万用表测电流如图 1-12 所示。

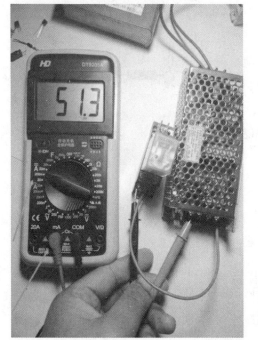
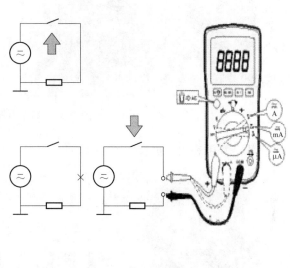

图 1-12　数字万用表测电流

1.2.4.3　电阻的测量

将表笔插进"COM"和"VΩ"孔中，把旋钮打旋到"Ω"中所需的量程，用表笔接在电阻两端金属部位，测量中可以用手接触电阻，但不要把手同时接触电阻两端，这样会影响测量精确度，人体是电阻很大但是有限大的导体。读数时，要保持表笔和电阻有良好的接触；注意单位：在"200"档时单位是"Ω"，在"2k"到"200k"档时单位为"kΩ"，"2M"以上的单位是"MΩ"。

需要注意的是：红表笔极性为"+"；开路显示为过量程状态，即显示"1"；测在线电阻时，须确认被测电路已关掉电源，同时电容已放电，方能进行测量。

数字万用表测电阻如图 1-13 所示。

1.2.4.4　二极管的测量

将量程开关置于二极管符号档位。黑表笔插入"COM"插孔，红表笔插入"VΩ"插孔。将表笔并接到被测二极管上，显示为正向压降的电压值，当二极管反接时，显示为过量程状态，输入端开路时，也显示为过量程状态，即最高位显示"1"。测试条件：正向直流电流约 1mA，反向直流电压约 3V。

· 141 ·

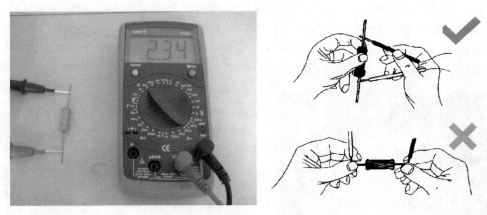

图 1-13　数字万用表测电阻

数字万用表测二极管如图 1-14 所示。

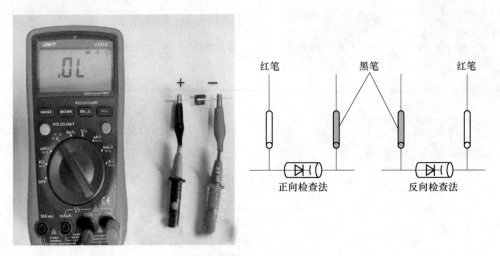

图 1-14　数字万用表测二极管

1.2.4.5　蜂鸣通断测试

黑表笔插入"COM"插孔，红表笔插入"V Ω"插孔。将量程开关置于蜂鸣器档位。将表笔跨接在预测线路的两端，当两点之间的电阻值小于 70Ω 时，蜂鸣器便会发出声响。

需要注意：当输入端开路时，仪表显示为过量程状态。被测电路必须在切断电源状态下检查通断，因为任何负载信号都可能会使蜂鸣器发声，导致错误判断。

1.2.4.6　晶体管的测量

表笔插位同上；其原理同二极管。先假定 A 脚为基极，用黑表笔与该脚相接，红表笔与其他两脚分别接触其他两脚；若两次读数均在 0.7V 左右，然后再用红笔接 A 脚，黑笔接触其他两脚，若均显示"1"，则 A 脚为基极，否则需要重新测量，且此管为 PNP 管。那么集电极和发射极如何判断呢？我们可以利用"h_{FE}"档来判断：先将档位打到"h_{FE}"

档，可以看到档位旁有一排小插孔，分为 PNP 和 NPN 管的测量。前面已经判断出管型，将基极插入对应管型"b"孔，其余两脚分别插入"c"和"e"孔，此时可以读取数值，即 β 值；再固定基极，其余两脚对调；比较两次读数，读数较大的引脚位置与表面"c"和"e"相对应。另外：上法只能直接对如 9000 系列的小型管测量，若要测量大管，可以采用接线法，即用小导线将 3 个引脚引出。

数字万用表测 h_{FE} 插孔示意图如图 1-15 所示。

图 1-15 数字万用表测 h_{FE} 插孔示意图

1.2.4.7 MOS 场效应晶体管的测量

N 沟道的有国产的 3D01、4D01，日产的 3SK 系列。G 极（栅极）的确定：利用万用表的二极管档。若某脚与其他两脚间的正反压降均大于 2V，即显示"1"，此脚即为栅极 G。再交换表笔测量其余两脚，压降小的，黑表笔接的是 D 极（漏极），红表笔接的是 S 极（源极）。

1.2.4.8 数字万用表的使用注意事项

(1) 如果开机后不显示任何数字，应首先检查 9V 集成电池是否已失效，还需检查电池引线有无断线，电池夹是否接触牢靠。若显示出低电压标志符，应及时更换新电池。测量时，倘若仅最高位显示数字"1"，其他位均消隐，证明仪表已发生过载，应选择更高的量程。有些数字万用表带读数保持开关或者按键，平时应置于关断位置，以免影响正常测量。一些新型数字万用表增加了自动关机功能，当仪表停止使用或停止于某一档位的时间超过 15min 时，能自动切断电源，使仪表处于低功耗的"休眠"状态，而并非出现故障。此时只需重新起动即可恢复正常工作。

(2) 使用数字万用表时不得超过所规定的极限值。最高 DC V 档的输入电压极限值为 1000V，最高 AC V 档则为 700V 或 750V（有效值）。当被测交流电压上叠加有直流电压时，两者电压之和不得超过所用 AC V 档的输入电压极限值。

（3）测量交流电时，应当用黑表笔接被测电压的低电位端（如被测信号源的公共地、机壳、220V 交流电的中性线端等），以消除仪表输入端对地分布电容的影响，减小测量误差。

（4）测量直流电压（或直流电流）时，仪表能自动判定并显示出电压（或电流）的极性，因此可不必考虑表笔的接法。

（5）测量大电流时须使用"20A"插孔，该插孔未加保护装置，因此测量大电流的时间不允许超过 10~15s，否则会影响读数的准确性。

（6）数字万用表的红表笔带正电，黑表笔带负电，这与指针式万用表电阻档的极性恰好相反。测量有极性的元器件时，必须注意表笔的极性。

（7）测量电阻时两手不得碰触表笔的金属端或元件的引出端，以免引入人体电阻，影响测量结果。严禁在被测线路带电的情况下测量电阻。

（8）用电容档测量电解电容器时，被测电容器的极性应与电容插座所标明的极性保持一致。测量之前必须将电容器放电，以免损坏仪表。

1.2.5 材料清单

材料清单见表 1-4。

表 1-4 材料清单

序号	名　　称	数量
1	XK-SX2C 型高级维修电工实训台	1 台
2	维修电工实训组件（XKDT11）	1 个
3	维修电工实训组件（XKDT12A）	1 个
4	数字万用表	1 个
5	电子元器件套件	1 套

1.2.6 任务实施

任务书见表 1-5。

表 1-5 任务书

（　　　　　　　　　　　　　　　　　　　　）任务书					
1. 根据要求完成所需器件清点并完成下表					
序号	器件名称	器件数量	器件型号	器件是否完好	备注

续表1-5

(　　　　　　　　　　　　　　　　) 任务书

1. 根据要求完成所需器件清点并完成下表

序号	器件名称	器件数量	器件型号	器件是否完好	备注

2. 项目任务描述

3. 任务所涉及的工作原理介绍

4. 数字万用表测量某元器件操作过程记录

续表 1-5

(				) 任务书	
1. 根据要求完成所需器件清点并完成下表					
序号	器件名称	器件数量	器件型号	器件是否完好	备注
5. 测量结果记录					
6. 任务小结					

1.2.7 任务评价

评分表见表 1-6。

表 1-6 评分表

(				) 评分表	
器件清点（10 分）					
序号	重点检查内容	评分标准	分值	得分	备注
1	器件清点	清点错误一项扣 1 分	5		
2	数字万用表功能核查	检查错误扣 1 分	5		
		小 计			
电流、电压的测量（20 分）					
序号	重点检查内容	评分标准	分值	得分	备注
1	电流的测量	错误一处扣 1 分	10		
2	电压的测量	错误一处扣 1 分	10		
		小 计			
元器件测量的工作原理描述（10 分）					
序号	重点检查内容	评分标准	分值	得分	备注
1	元器件的特性	是否完整酌情	5		
2	工作原理	是否完整酌情	5		
		小 计			

续表1-6

( ) 评分表

数字万用表测量元器件的步骤 (30分)

序号	重点检查内容	评分标准	分值	得分	备注
1	表笔的接线	错误一处扣1分	10		
2	档位的选择	错误一处扣1分	10		
3	元器件的测量位置	错误一处扣1分	10		
		小 计			

测量结果 (20分)

序号	重点检查内容	评分标准	分值	得分	备注
1	是否会看测量数据		10		
2	是否能判定元器件是否损坏		10		
		小 计			

职业素养 (10分)

序号	重点检查内容	评分标准	扣分	得分	备注
1	带电操作	一次扣5分扣完为止			
2	规范操作	一次扣1分扣完为止			
3	团队合作	酌情			
4	工位整洁	酌情			
	总 计				

1.2.8 任务拓展

数字万用表常见故障及检修

(1) 常见故障。寻找故障应先外后里、先易后难、化整为零、重点突破。其方法大致可分为以下几种：

1) 感觉法。凭借感官直接对故障原因做出判断，通过外观检查，能发现如断线、脱焊、搭线短路、熔丝管断、烧坏元器件、机械性损伤、印制电路上铜箔翘起及断裂等故障；可以触摸出电池、电阻、晶体管、集成块的温升情况，可参照电路图找出温升异常的原因。另外，用手还可检查元器件有否松动、集成电路引脚是否插牢，转换开关是否卡带；可以听到和嗅到有无异声、异味。

2) 测电压法。测量各关键点的工作电压是否正常，可较快找出故障点。如测A-D转换器的工作电压、基准电压等。

3) 短路法。在前面所讲的检查A-D转换器方法里一般都采用短路法，这种方法在修理弱电和微电仪器时用得较多。

4) 断路法。把可疑部分从整机或单元电路中断开，若故障消失，表示故障在断开的电路中。此法主要适合于电路存在短路的情况。

5) 测元器件法。当故障已缩小到某处或几个元器件时，可对其进行在线或离线测量。必要时，用好的元器件进行替换，若故障消失，说明该元器件已坏。

6) 干扰法。利用人体感应电压作为干扰信号，观察液晶显示的变化情况，常用于检查输入电路与显示部分是否完好。

(2) 故障解决方法。对一块故障仪表首先应检查和判别故障现象是共性（所有功能都不能测量），还是个性（个别功能或个别量程），然后区别情况，对症解决。

若所有档均不能工作，应重点检查电源电路和 A-D 转换器电路。检查电源部分时，可取下叠层电池，按下电源开关，用正表笔接被测表电源负，负表笔接电源正（对数字万用表而言），开关打到二极管测量档，若显示的是二极管正向电压，则说明电源部分是好的，若偏差大，则说明电源部分有问题。若出现开路，重点检查电源开关和电池引线等。若出现短路，则需要采用断路法，逐步断开使用电源的元器件，重点检查运算放大器、定时器及 A-D 转换器等。若出现短路，一般都不止损坏一块集成元器件。检查 A-D 转换器可以和基本表同时进行，相当于模拟式万用表的直流表头。具体检查方法：首先将被测表的量程转到直流电压最低档；其次测量 A-D 转换器工作电压是否正常。根据表内所用A-D 转换器型号，对应 V_+ 脚和 COM 脚，测量值与它的典型值相比较是否相符。再次，测A-D 转换器的基准电压，目前常用的数字万用表的基准电压一般都是 100mV 或 1V，即测量 U_{REF+} 与 COM 之间的直流电压，若偏离 100mV 或 1V，可通过外接电位器进行调节。检查输入为零的显示数，把 A-D 转换器的正端 IN_+ 与负端 IN_- 短接，使输入电压 $U_{in}=0$，仪表显示 "00.0" 或 "00.00"。检查显示器的全亮笔划。把测试端 TEST 脚与正电源端 V_+ 短接，使逻辑地变成高电位，全部数字电路停止工作。因每个笔划上都加有直流电压，所以全部笔划亮对位表显示 "1888"，对位表显示 "18888"。若存在缺笔划现象，检查 A-D 转换器对应输出脚与导电胶（或联线），与显示器之间是否有接触不良和断线情况。

若个别档有问题，说明 A-D 转换器和电源部分都工作正常。因直流电压、电阻档共用一套分压电阻；交直流电流共用分流器；交流电压与交流电流共用一套 AC-DC 转换器；其他如 Cx、h_{FE}、F 等都由独立的不同转换器组成。了解它们之间的关系，再根据电源图，就很容易找到故障部位。若测量小信号不准确或显示数字跳动大，则重点检查量程开关的接触是否良好。

若出现测量数据不稳，且数值总是累计增大，短接 A-D 转换器的输入端，显示数据不为零的情况，则一般是 0.1μF 的基准电容性能不良所引起的。

根据以上分析，数字万用表的修理基本顺序应是：数字表头部→直流电压→直流电流→交流电压→交流电流→电阻档（包括蜂鸣器和检查二极管正压降）→Cx→h_{FE}、F、H、T 等。但也不可过分机械，有些明显能看出的问题，可以先处理。但在进行调校时，则一定要按照上述程序。

总之，一块故障万用表，经过适当的检测，首先要分析故障可能出现的部位，然后根据线路图找到故障位置进行更换和修复。因数字万用表是较精密的仪表，更换元器件一定要用参数相同的元器件，特别是更换 A-D 转换器，一定要采用生产厂家经严格筛选的集成块，否则将出现误差而达不到所需准确度。新换的 A-D 转换器，也需要按前面所述的方法进行检查，切不可因新而置信不疑。

目前，国内生产数字万用表的厂家甚多，质量也有优劣，对双面复铜板的质量问题，在修理中是不易发现的。树脂板的绝缘强度不够时，主要表现在测量高电压时误差较大，修理时要与分压电阻的阻值变化区别开来。遇到这种情况，最好是采用断路法，寻找故障点。对烧坏碳化的部分要清除干净，达到绝缘要求。遇到由双面连线因过渡孔断裂而引起不能输入信号时，容易与转换开关不良的现象混淆而难以分开，这类故障宜采用短路法寻找故障点。

故障排除记录单见表 1-7。

表 1-7 故障排除记录单

		故障排除记录单
故障1	故障现象	
	故障原因	
	故障的排除过程	
故障2	故障现象	
	故障原因	
	故障的排除过程	
故障3	故障现象	
	故障原因	
	故障的排除过程	
故障4	故障现象	
	故障原因	
	故障的排除过程	

1.2.9 任务小结

数字万用表在测量电阻时，无论被测对象是复杂的系统还是单个的电子元器件，被测部分不得存在除数字万用表测试电流以外的任何电源所形成的电流成分，而且测量者的手指等身体的任何部位均不得接触两个表笔的触针及被测对象的导电部位。在测量较高的电压和较大的电流的过程中，不要切换选择开关的档位，否则很容易烧伤开关触点及损坏仪表内部的其他电子元器件。如果情况特殊而确需切换时，应将表笔离开被测量电路，再进行切换。当数字万用表处在二极管测试档、蜂鸣器档和电阻档时，红表笔连接着仪表内部高电位而带正电，黑表笔内接虚地而带负电。

普通数字万用表的交流电压档属于平均值仪表，而且是按正弦波特性设计的，所以它不能直接用于测量锯齿波、三角波、矩形波等非正弦波电压，被测量即便是正弦波电压，如果其波形的失真度较大，仍将不能获得正确的测量结果。我们在进行测量时，这些问题都应该注意。

项目 2　三相异步电动机的基本控制技能训练

任务 2.1　三相鼠笼式异步电动机的认识与基本测试

2.1.1　任务描述

某工厂一线员工小王，在岗位工作中，手背偶然触碰到了正在运行的电动机的外壳，身体感觉发麻，急忙将手撤回，之后小王关停电动机，将设备断电，对电动机进行检测、维修。

大家思考一下，小王为什么身体感觉发麻？当身体触碰到电动机身体发麻这种现象正常吗？

本任务针对上述问题，学习三相鼠笼异步电动机的结构、工作原理，以及通过铭牌认识电动机的性能。

此外，在实际使用中，还要保证电动机的外壳、绕组之间绝缘性良好，以免发生安全问题，如何进行测试呢？

2.1.2　任务目标

(1) 熟悉三相鼠笼式异步电动机的结构。
(2) 掌握三相鼠笼式异步电动机的工作原理。
(3) 掌握能看懂三相鼠笼式异步电动机的铭牌。
(4) 掌握检验三相异步电动机绝缘情况的方法。
(5) 三相异步电动机的拆卸和安装。
(6) 掌握安全操作注意事项。

2.1.3　任务分析

在任务描述中员工小王偶然触碰到运行中的电动机，身体感觉到发麻，是因为电动机外壳绝缘性能变差，性能良好的电动机外壳绝缘性要好，不能带电。这就需要我们掌握检测电动机外壳绝缘性的方法，掌握电动机的结构、工作原理等知识技能，以便对电动机拆装维修。

在生产实际中，电动机是生产机械的原动机，其作用是将电能转换成机械能。机床、起重机、锻压机、鼓风机、水泵，大多数生产机械都用它来驱动。电动机按照转子结构的不同，可分为鼠笼式电动机和绕线式电动机。三相鼠笼式异步电动机具有结构简单、运行可靠、价格便宜、使用、安装、维护方便等优点，被广泛应用于各个领域。三相异步电动机的外形如图 2-1 所示。

本任务分为：

(1) 三相鼠笼式异步电动机的结构；
(2) 三相异步电动机的工作原理；

图 2-1　三相鼠笼式异步电动机的外形

(3) 三相异步电动机的铭牌数据；
(4) 三相鼠笼式异步电动机的检查。

2.1.4　任务相关知识

2.1.4.1　三相鼠笼式异步电动机的结构

(1) 定子部分。三相异步电动机定子作用是产生旋转磁场。定子主要由机座、定子铁心、定子绕组等组成。定子铁心是电动机磁路的一部分，它由互相绝缘的硅钢片叠成圆筒形，装在机座内壁上。在定子铁心的内圆周表面均匀冲有槽孔，用以嵌放定子绕组。定子铁心如图 2-2 所示，定子绕组如图 2-3 所示。

图 2-2　定子铁心

图 2-3　定子绕组

(2) 转子部分。三相异步电动机转子作用是在定子磁场作用下，产生感应电动势或电流，转动输出机械能。转子主要由转轴、转子铁心和转子绕组等组成。转轴上压装着由硅钢片叠成的圆柱形转子铁心，转子铁心也是电动机磁路的一部分，在转子铁心的外圆周表面均匀冲有槽孔，槽内嵌放转子绕组。分为笼型与绕线型。笼型转子铁心与绕组如图 2-4 所示，绕线型转子铁心与绕组如图 2-5 所示。

图 2-4　笼型转子铁心与绕组

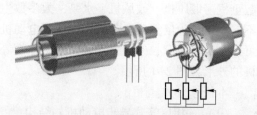

图 2-5　绕线型转子铁心与绕组

(3) 三相异步电动机的结构拆分。

三相异步电动机的结构拆分图如图 2-6 所示。

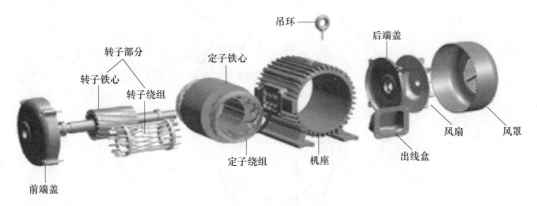

图 2-6 三相异步电动机的结构拆分图

2.1.4.2 三相异步电动机的工作原理

(1) 旋转磁场的产生。三相对称交流电（见图 2-7）通入三相定子绕组（三个线圈彼此互隔 120°分布在定子铁心内圆的圆周上）。三相异步电动机定子绕组通入三相对称交流电，示意图如图 2-8 所示。

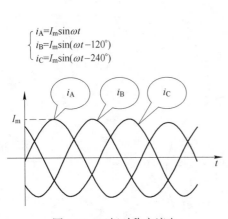

图 2-7 三相对称交流电

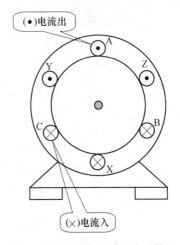

图 2-8 三相异步电动机定子绕组通入三相对称交流电

通过分析不同时间产生的磁场的位置变化，可知三相交流电通入三相定子绕组会产生旋转磁场。旋转磁场的产生如图 2-9 所示。

(2) 转子转动的原理。转子转动的原理如图 2-10 所示。

三相异步电动机的工作原理：三相交流电通入三相异步电动机的定子绕组，产生旋转磁场，静止的转子相对于旋转磁场有一个相对的切割磁力线的运动，产生感应电动势，并产生感应电流，转子绕组上有了电流，在磁场中会受到电磁力的作用，形成电磁转矩 T，

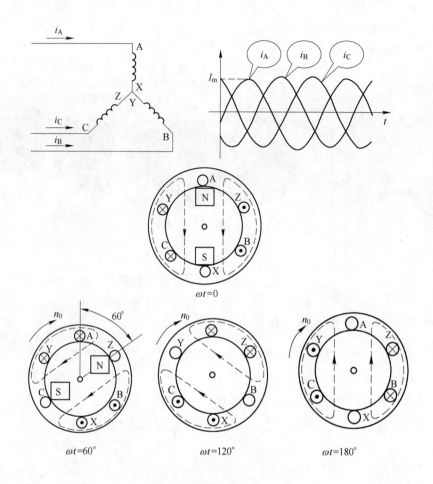

图 2-9 旋转磁场的产生

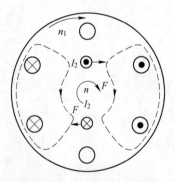

图 2-10 转子转动的原理

克服阻转矩,驱动转子旋转起来,实现了电能转换成机械能的目的。

(3) 旋转磁场的旋转方向。旋转磁场的旋转方向取决于三相电流的相序。改变流入三相绕组的电流相序,就能改变旋转磁场的转向;改变了旋转磁场的转向,也就改变了三相异步电动机的旋转方向。三相电流的相序对三相异步电动机转向的影响如图 2-11 所示。

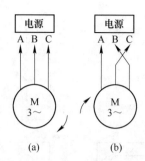

图 2-11 三相电流的相序对三相异步电动机转向的影响
(a) 正转；(b) 反转

2.1.4.3 三相异步电动机的铭牌数据

(1) 型号。型号是不同种类和型式电动机的代号，它的每一个字母都具有一定的含义。常用的 Y 系列异步电动机有 Y (IP44) 封闭式、Y (IP23) 防护式小型三相异步电动机，YR (IP44) 封闭式、YR (IP23) 防护式绕线型三相异步电动机，YD 变极多速三相异步电动机，YX 高效率三相异步电动机，YH 高转差率三相异步电动机，YB 隔爆型三相异步电动机，YCT 电磁调速三相异步电动机，YEJ 制动三相异步电动机，YTD 电梯用三相异步电动机，YQ 高起动转矩三相异步电动机等几十种产品。Y 系列电动机具有高效、节能、特性好、噪声低等优点，功率等级和安装尺寸符合国际标准。三相异步电动机的型号（例如 Y132M-4）如下所示。

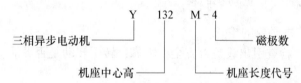

(2) 额定容量（功率）。额定容量（功率）指电动机在额定运行时，电动机转轴上输出的功率。

(3) 频率。频率指加在电动机定子绕组上的允许频率。

(4) 额定电压。额定电压指定子绕组在指定接法下应加的线电压。

(5) 额定电流。额定电流指定子绕组在指定接法下的线电流。

(6) 联结方式。三相负载有 Y、△ 两种联结方法。三相异步电动机的定子绕组两种联结方法如图 2-12 所示。

(7) 转速。电机轴上的转速 (n)。

(8) 绝缘等级。绝缘等级是由电动机所用的绝缘材料决定的，按耐热程度不同，绝缘材料分为 A、E、B、F、H 等数级。

(9) 工作方式。电动机按持续运行的时间设计工作方式，分为连续（S_1）、短时（S_2）和重复短时（S_3）三种工作方式。

连续工作方式表示这种电动机可以按铭牌上规定的功率长期连续运行。

短时工作方式表示这种电动机不能连续运行，在额定功率输出时，只能按铭牌上的规定短时运行。

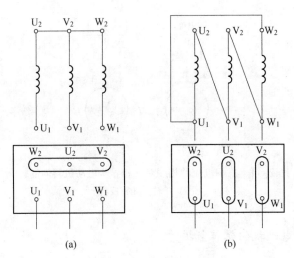

图 2-12 三相异步电动机的定子绕组两种联结方法
(a) Y 接法；(b) △ 接法

重复短时工作方式表示这种电动机不能在额定功率输出时连续运行，只能按规定时间做重复性短时运行。

(10) 功率因数（$\cos\varphi 1$）。指电动机在额定运行时的功率因数。

(11) 效率。效率指电动机输出功率与输入功率的比值。

2.1.4.4 三相鼠笼式异步电动机的检查

电动机使用前应做必要的检查。

(1) 机械检查。检查引出线是否齐全、牢靠；转子转动是否灵活、匀称，是否有异常声响等。

(2) 电气检查。电动机在日常运行中常会有线圈松动，使绝缘磨损老化，或表面受污染、受潮等引起绝缘电阻日趋下降，绝缘电阻降低到一定值会影响电动机起动和正常运行，甚至会损坏电动机，危及人身安全。因此在各类电动机开始使用之前或经过霉季、受潮、重新安装之后，首先要测定各相绕组对机壳的绝缘电阻及绕组之间的绝缘电阻。绝缘电阻的测量一般用绝缘电阻表进行，学会绝缘电阻表的使用，在检查电机、电器及线路的绝缘情况和测量高值电阻时能给我们带来方便。

绝缘电阻表是一种简便常用的测量高电阻的仪表，主要用来检查电气设备、家用电器或电气线路对地及相间的绝缘电阻，以保证这些设备、电器和线路工作在正常状态，避免发生触电伤亡及设备损坏等事故。绝缘电阻表大多采用手摇发电机供电，它的刻度是以兆欧（$M\Omega$）为单位的。

常见的绝缘电阻表主要由作为电源的高压手摇发电机和磁电式流比计两部分组成，绝缘电阻表的结构示意图及其原理电路图如图 2-13 所示。

1) 绝缘电阻表的工作原理。与绝缘电阻表表针相连的有两个线圈，一个同表内的附加电阻 R 串联；另一个和被测电阻 R_x 串联，然后一起接到手摇发电机上。当手摇动发电机时，两个线圈中同时有电流通过，在两个线圈上产生方向相反的转矩，表针就随着两个

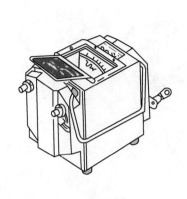
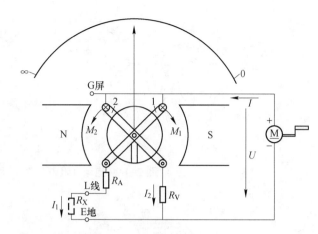

图 2-13 绝缘电阻表外观及电路原理图

转矩的合成转矩的大小而偏转某一角度,这个偏转角度决定于两个电流的比值,附加电阻是不变的,所以电流值仅取决于待测电阻的大小。

2) 绝缘电阻表的选择。在测量电气设备的绝缘电阻之前,先要根据被测设备的性质和电压等级,选择适当的绝缘电阻表。

一般测量额定电压在 500V 以下的设备时,选用 500~1000V 的绝缘电阻表,测量额定电压在 500V 以上的设备时,选用 1000~2500V 的绝缘电阻表。例如,测量高压设备的绝缘电阻,不能用额定电压 500V 以下的绝缘电阻表,因为这时测量结果不能反映工作电压下的绝缘电阻;同样不能用电压太高的绝缘电阻表测量低压电气设备的绝缘电阻,否则会损坏设备的绝缘。

此外,绝缘电阻表的测量范围也应与被测绝缘电阻的范围相吻合。一般应注意不要使其测量范围过多地超出所需测量的绝缘电阻值,以免使读数产生较大误差。一般测量低压电气设备绝缘电阻时,可选用 0~200MΩ 量程的表,测量高压电气设备或电缆时可选用 0~2000MΩ 量程的表。刻度不是从零开始,而是从 1MΩ 开始的,绝缘电阻表一般不宜用来测量低压电气设备的绝缘电阻。

3) 使用前的检查。绝缘电阻表使用前要先进行一次开路和短路试验,检查绝缘电阻表是否良好。

将"L""E"端开路,摇动手柄,指针应指在"∞"处,再将"L""E"端短接,摇动手柄,指针应指在"0"处,说明绝缘电阻表是良好的,否则是有误差的。

绝缘电阻表开路和短路试验如图 2-14 所示。

4) 绝缘电阻表的接线。绝缘电阻表上有 3 个分别标有接地"E"、电路"L"和保护环"G"的接线柱。

①测量电路绝缘电阻时,可将被测端接于电路"L"接线柱上,以良好的地线接于接地"E"的接线柱上,如图 2-15(a)所示。

②测量电动机绝缘电阻时,将电动机绕组接于电路"L"端,机壳接于接地"E"端,如图 2-15(b)所示。

③测量电动机绕组间的绝缘性能,将电路"L"端和接地"E"端分别接在电动机两绕组的接线端。

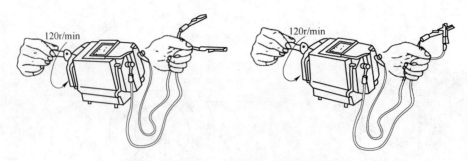

图 2-14 绝缘电阻表开路和短路试验

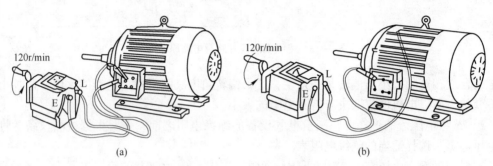

图 2-15 用绝缘电阻表检查电机绕组间及绕组与机壳之间的绝缘性能
(a) 电机绕组之间的绝缘检查；(b) 电机绕组与机壳之间的绝缘检查

④测量电缆的缆芯对缆壳的绝缘电阻时，除将缆芯接于电路"L"端和缆壳接于接地"E"端以外，还要将电缆壳、芯之间的内层绝缘物接保护环"G"，以消除因表面漏电而引起的误差。

5) 使用绝缘电阻表的注意事项：

①在进行测量前要先切断电源，被测设备一定要进行放电（约需 2~3min），以保障设备及人身安全。

②接线柱与被测设备间连接的导线不能用双股绝缘线或绞线，应用单股线分开单独连接，避免因绞线绝缘不良引起误差，应保持设备表面清洁干燥。

③测量时，表面应放置平稳，手柄摇动要由慢渐快。

④一般采用均匀摇动 1min 后的指针位置作为读数。一般为 120r/min。测量中如发现指针指示为 0，则应停止转动手柄，以防表内线圈过热而烧坏。

⑤在绝缘电阻表转动尚未停下或被测设备未放电时，不可用手进行拆线，以免引起触电。

2.1.5 材料清单

材料清单见表 2-1。

表 2-1 材料清单

序号	名 称	数量
1	XK-SX2C 型高级维修电工实训台	1 台

续表 2-1

序号	名　　称	数量
2	三相异步电动机	1台
3	跨接线	若干
4	ZC25-4型绝缘电阻表	若干
5	万用表	若干
6	工作服、绝缘鞋、安全帽等	若干
7	螺钉旋具、扳手、套管、老虎钳	若干

2.1.6　任务实施

任务实施见表2-2。

表 2-2　任务实施

三相鼠笼式异步电动机的认识与基本测试任务书					
1. 根据要求完成所需器件清点并完成下表					
序号	器件名称	器件数量	器件参数	器件测量	备注
2. 电动机工作原理描述					

续表 2-2

三相鼠笼式异步电动机的认识与基本测试任务书

3. 电动机拆装过程记录

4. 电动机绝缘性能检测过程记录

5. 任务小结

2.1.7 任务评价

任务评价见表 2-3。

表 2-3 任务评价

电动机拆装及检测评分表					
器件清点和测量（10分）					
序号	重点检查内容	评分标准	分值	得分	备注
1	器件清点	器件清点错误一项扣1分	5		
2	仪表使用，档位选择	使用、档位错误一项扣1分	5		
小　计					

续表 2-3

电动机拆装及检测评分表

电动机拆卸（30分）

序号	重点检查内容	评分标准	分值	得分	备注
1	拆卸无误，摆放整齐	错误一处扣2分	10		
2	无损伤电动机器件	损伤一处扣10分	20		
	小 计				

电动机绝缘检测（20分）

序号	重点检查内容	评分标准	分值	得分	备注
1	绝缘电阻表使用前设置正确	错误一处扣5分	10		
2	绝缘电阻表使用正确	错误一处扣5分	10		
	小 计				

电动机组装（30分）

序号	重点检查内容	评分标准	分值	得分	备注
1	安装顺序是否正确	错误一处扣5分	10		
2	是否损伤器件	错误一处扣5分	10		
3	组装是否牢固	错误一处扣5分	10		
	小 计				

职业素养（10分）

序号	重点检查内容	评分标准	扣分	得分	备注
1	带电操作	一次扣5分扣完为止			
2	规范操作	一次扣1分扣完为止			
3	团队合作	酌情			
4	工位整洁	酌情			
	总 计				

2.1.8 故障的排除

（1）电动机外壳带电。一般要求电动机泄漏电流不应大于0.8mA，以保证人身安全。电动机外壳漏电的主要原因有电动机内某引出线绝缘破损并碰触壳体；电动机绕组局部烧毁引起定子与外壳间漏电。较多见的是长期处于高湿环境，导致电动机受潮绝缘能力降低而使壳体带电。此时，可用绝缘电阻表测量电动机各绕组与机壳间的绝缘电阻值，若在2MΩ以下，则说明电动机已受潮严重，应将电动机定子绕组进行烘烤去潮处理。

（2）定、转子铁心故障检修。定、转子都是由相互绝缘的硅钢片叠成，是电动机的磁路部分。定、转子铁心的损坏和变形主要由以下几个方面原因造成。

1）轴承过度磨损或装配不良，造成定、转子相擦，使铁心表面损伤，进而造成硅钢片间短路，电动机铁损增加，使电动机温升过高。这时应用细锉等工具去除毛刺，消除硅钢片短接，清理干净后涂上绝缘漆，并加热烘干。

2) 拆除旧绕组时用力过大，使齿槽歪斜和向外张开。此时应用尖嘴钳、木榔头等工具予以修整，使齿槽复位，并在不好复位的有缝隙的硅钢片间加入青壳纸、胶木板等硬质绝缘材料。

3) 因受潮等原因造成铁心表面锈蚀。此时需用砂纸打磨干净，清理后涂上绝缘漆。

4) 围绕组接地产生高热烧毁铁心槽或齿部。可用凿子或刮刀等工具将熔积物剔除干净，涂上绝缘漆烘干。

5) 铁心与机座间结合松动，可拧紧原有定位螺钉。若定位螺钉失效，可在机座上重钻定位孔并轻敲机座，旋紧定位螺钉。

(3) 轴承故障检修。转轴通过轴承支撑转动，是负荷最重的部分，又是容易磨损的部件。

1) 故障检查。

①运行中检查：滚动轴承缺油时，会听到"骨碌骨碌"的声音；若听到不连续的"梗梗"的声，可能是轴承钢圈破裂。轴承内混有沙土等杂物或轴承零件有轻度磨损时，会产生轻微的杂音。

②拆卸后检查：先查看轴承滚动体、内外钢圈是否有破损、锈蚀、疤痕等，然后用手捏住轴承内圈，并使轴承摆平，另一只手用力推外钢圈，如果轴承良好，外钢圈应转动平稳，转动中无振动和明显的卡滞现象，停转后外钢圈没有倒退现象。否则说明轴承已不能再用了。

左手卡住外圈，右手捏住内钢圈，用力向各个方向推动，如果推动时感到很松，就是磨损严重。

2) 故障修理。轴承外表面上的锈斑可用00号砂纸擦除，然后放入汽油中清洗；或轴承有裂纹、内外圈碎裂或轴承过度磨损时，应更换新轴承。更换新轴承时，要选用与原来型号相同的轴承。

(4) 转轴故障检修：

1) 轴弯曲。若弯曲不大，可通过磨光轴颈、滑环的方法进行修复；若弯曲超过0.2mm，可将转轴放于压力机下，在拍弯曲处加压矫正，矫正后的轴表面用车床切削磨光；如果弯曲过大，则需另换新轴。

2) 轴颈磨损。轴颈磨损不大时，可在轴颈上镀一层铬，再磨削至需要尺寸；磨损较多时，可在轴颈上进行堆焊，再到车床上切削磨光；如果轴颈磨损过大时，也在轴颈上车削2~3mm，再车一套筒，趁热套在轴颈上，然后车削到所需尺寸。

3) 轴裂纹或断裂。轴的横向裂纹深度不超过轴直径的10%~15%、纵向裂纹不超过轴长的10%时，可用堆焊法补救，然后再精车至所需尺寸。若轴的裂纹较严重，就需要更换新轴。

(5) 机壳和端盖的检修。机壳和端盖若有裂纹应进行堆焊修复。若遇到轴承镗孔间隙过大，造成轴承端盖配合过松，一般可用冲子将轴承孔壁均匀打出毛刺，然后再将轴承打入端盖。对于功率较大的电动机，也可采用镶补或电镀的方法，最后加工出轴承所需要的尺寸。

故障排除记录单见表2-4。

表 2-4 故障排除记录单

		故障排除记录单
故障 1	故障现象	
	故障原因	
	故障的排除过程	
故障 2	故障现象	
	故障原因	
	故障的排除过程	
故障 3	故障现象	
	故障原因	
	故障的排除过程	
故障 4	故障现象	
	故障原因	
	故障的排除过程	

2.1.9 任务拓展

电动机的拆装，三相异步电动机的结构拆分图如图 2-6 所示。

（1）三相异步电动机的一般拆卸步骤：

1）切断电源，卸下皮带。

2) 拆去接线盒内的电源接线和接地线。

3) 卸下底脚螺母、弹簧垫圈和平垫片。

4) 卸下皮带轮。

5) 卸下前轴承外盖。

6) 卸下前端盖。可用大小适宜的扁凿,插在端盖突出的耳朵处,按端盖对角线依次向外撬,直至卸下前端盖。

7) 卸下风叶罩和卸下风叶。

8) 卸下后轴承外盖。

9) 卸下后端盖和转子。

在抽出转子之前,应在转子下面和定子绕组端部之间垫上厚纸板,以免抽出转子时碰伤定子铁心和绕组。

10) 最后用拉具拆卸前后轴承及轴承内盖。

(2) 三相异步电动机的一般装配步骤:电动机的装配顺序按拆卸的逆序进行。

1) 安装轴承。

2) 安装转子。

3) 安装端盖。

4) 安装轴承外盖。

5) 安装风扇和风罩。

6) 皮带轮或联轴器的安装。

7) 装配后的一般检验。

(3) 安装轴承的几种方法。安装前的准备工作:

1) 将轴承和轴承盖用煤油清洗后,检查轴承有无裂纹,滚道内有无锈迹等。

2) 再用手旋转轴承外圈,观察其转动是否灵活、均匀。来决定轴承是否要更换。

3) 如不需要更换,再将轴承用汽油洗干净,用清洁的布擦干待装。更换新轴承时,应将其放在70~80℃的变压器油中,加热5min左右,待全部防锈油溶去后,再用汽油洗净,用洁净的布擦干待装。

几种常用的安装方法:

(1) 敲打法。把轴承套到轴上,对准轴颈,用一段铁管,其内径略大于轴颈直径,外径略大于轴承内圈的外径,铁管的一端顶在轴承的内圈上,用手锤敲铁管的另一端,把轴承敲进去。如果没有铁管,也可用铁条顶住轴承的内圈,对称地、轻轻地敲,轴承也能水平地套入转轴。

(2) 热装法。如配合度较紧,为了避免把轴承内环胀裂或损伤配合面,可采用热装法。首先将轴承放在油锅(或油槽内)里加热,油的温度保持在100℃左右,轴承必须浸没在油中,又不能和锅底接触,可用铁丝将轴承吊起架空。加热要均匀,30~40min后,把轴承取出,趁热迅速地将轴承一直推到轴颈。在农村可将轴承放在100W灯泡上烤热,1h后即可套在轴上。

2.1.10 任务小结

性能良好的电动机机壳绝缘性能良好,当机壳绝缘性变差时,会产生漏电,这时,就需要对电动机断电，进行检测及维修。

检测的方法是利用绝缘电阻表检测绕组之间是否绝缘，机壳与绕组之间是否绝缘。如果绕组之间或绕组与机壳之间阻值不为无穷大，则电动机存在故障。

电动机外壳漏电的主要原因有电动机内某引出线绝缘破损并碰触壳体；电机绕组局部烧毁引起定子与外壳间漏电。此时，需要拆卸维修内部电路。

如果是处于高湿环境，导致电动机受潮绝缘能力降低而使机壳带电。应将电动机定子绕组进行烘烤去潮处理。

本任务主要学习了三相鼠笼式异步电动机的结构；三相异步电动机的工作原理；三相异步电动机的铭牌数据；三相鼠笼式异步电动机的检查；三相异步电动机的拆装等内容，为后期学习三相异步电动机电路的设计及安装打下了基础。

任务2.2　三相异步电动机的起动电路安装调试

2.2.1　任务描述

在工农业生产中，都要用到电动机，由电动机带动生产设备的运行，将电能转化为机械能。那么，工农生产中如何控制电动机起动运行呢？要用到哪些辅助器件呢？这些辅助器件的结构、工作原理是什么？如何使用呢？

2.2.2　任务目标

(1) 熟悉各种低压电器的结构、工作原理。
(2) 掌握各种低压电器的使用。
(3) 掌握三相异步电动机直接起动控制电路的接线、查线和操作。
(4) 掌握三相异步电动机点动、自锁起动、点动与长动控制电路原理。
(5) 掌握短路保护、过载保护、失压保护、欠压保护的原理。

2.2.3　任务分析

电动机接通电源后由静止状态逐渐加速到稳定状态的过程称为电动机的起动。

直接起动又称为全压起动，它是通过开关或接触器将额定电压直接加在电动机的定子绕组上而使电动机起动的方法。

控制电动机的起动，就要运用到各种低压电器，主要有断路器、按钮、交流接触器、热继电器、熔断器等。

电动机的起动控制线路主要有手动控制线路、点动控制线路、自锁控制线路、过载保护的控制起动线路和既能连续运行又能点动控制的控制线路等。

这些控制线路各式各样，但是不管是简单还是复杂，一般都是由一些基本控制环节组成的，在分析控制线路的原理和判断其故障时，一般都是从这些基本控制环节着手。

因此，从学习各式各样低压电器开始，到分析由其组成的基本控制线路，对生产机械整个电气控制线路的工作原理分析及维修有着很大的帮助。

本任务主要学习各种低压电器和电动机的手动控制线路、点动控制线路、自锁控制线路、过载保护的控制起动线路和既能连续运行又能点动控制的控制线路。

本任务分为：

(1) 认识低压电器；

(2) 三相异步电动机手动正转控制线路；

(3) 三相异步电动机点动控制线路；

(4) 三相异步电动机自锁控制线路；

(5) 三相异步电动机过载保护控制线路。

2.2.4 任务相关知识

2.2.4.1 认识低压电器

凡是对电能的生产、输送、分配和使用起控制、调节、检测、转换及保护作用的电工器械均可称为电器。用于交流1200V以下、直流1500V以下电路，起通断、控制、保护与调节等作用的电器称为低压电器。

(1) 低压电器的分类。低压电器的功能多、用途广、品种规格繁多，为了系统地掌握，必须加以分类。

1) 按电器的动作性质分。

①手动电器：人操作发出动作指令的电器，例如刀开关、按钮等。

②自动电器：不需人工直接操作，按照电的或非电的信号自动完成接通、分断电路任务的电器，例如接触器、继电器、电磁阀等。

2) 按用途分。

①控制电器：用于各种控制电路和控制系统的电器，例如接触器、继电器、电动机起动器等。

②配电电器：用于电能的输送和分配的电器，例如刀开关、低压断路器等。

③主令电器：用于自动控制系统中发送动作指令的电器，例如按钮、转换开关等。

④保护电器：用于保护电路及用电设备的电器，例如熔断器、热继电器等。

⑤执行电器：用于完成某种动作或传送功能的电器，例如电磁铁、电磁离合器等。

3) 按工作原理分。

①电磁式电器：依据电磁感应原理来工作的电器，如交直流接触器、各种电磁式继电器等。

②非电量控制电器：电器的工作是靠外力或某种非电物理量的变化而动作的电器，如刀开关、速度继电器、压力继电器、温度继电器等。

(2) 断路器。断路器是指能够关合、承载和开断正常回路条件下的电流并能关合、在规定的时间内承载和开断异常回路条件下的电流的开关装置。断路器按其使用范围分为高压断路器与低压断路器，高低压界线划分比较模糊，一般将3kV以上的称为高压电器。

断路器可用来分配电能，不频繁地启动异步电动机，对电源线路及电动机等实行保护，当它们发生严重的过载或者短路及欠压等故障时能自动切断电路，其功能相当于熔断器式开关与过欠热继电器等的组合。而且在分断故障电流后一般不需要变更零部件。目前，已获得了广泛应用。

断路器一般由触头系统、灭弧系统、操作机构、脱扣器、外壳等构成。

小型断路器如图 2-16 所示。小型断路器俗称空气开关，是塑料外壳式断路器的一种。是建筑电气终端配电装置引中使用最广泛的一种终端保护电器。用于 125A 以下的单相、三相的短路、过载、过压等保护，包括单极 1P、二极 2P、三极 3P、四极 4P 四种。

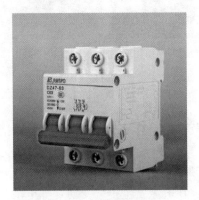

图 2-16　小型断路器

（3）熔断器。熔断器是指当电流超过规定值时，以本身产生的热量使熔体熔断，断开电路的一种电器。熔断器是根据电流超过规定值一段时间后，以其自身产生的热量使熔体熔化，从而使电路断开；运用这种原理制成的一种电流保护器。熔断器广泛应用于高低压配电系统和控制系统以及用电设备中，作为短路和过电流的保护器，是应用最普遍的保护器件之一。

1）熔断器的结构和作用。熔断器的结构一般分成熔体座和熔体等部分。熔断器是串联连接在被保护电路中的，当电路电流超过一定值时，熔体因发热而熔断，使电路被切断，从而起到保护作用。熔体的热量与通过熔体电流的平方及持续通电时间成正比，当电路短路时，电流很大，熔体急剧升温，立即熔断，当电路中电流值等于熔体额定电流时，熔体不会熔断。所以熔断器可用于短路保护。由于熔体在用电设备过载时所通过的过载电流能积累热量，当用电设备连续过载一定时间后熔体积累的热量也能使其熔断，所以熔断器也可作过载保护。熔断器的电气符号如图 2-17 所示，常用的熔断器外形如图 2-18 所示。

图 2-17　熔断器电气符号

其中，螺旋式熔断器和瓷插式熔断器是工业生产中常用的熔断器，两者的作用相同，用于电气设备的过载及短路保护。分断能力较强、结构紧凑、体积小、安装面积小、更换熔体方便、工作安全可靠。

  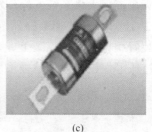 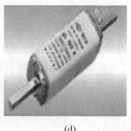

(a) (b) (c) (d)

图 2-18 熔断器外形图

(a) 瓷插式；(b) 螺旋式；(c) 无填料密封管式；(d) 有填料密封管式

熔断管内装有熔丝和石英砂，上有熔断指示器，指示器指示熔丝是否熔断。石英用于增强灭弧性能。

2) 熔断器的选择。对熔断器的要求是：在电气设备正常运行时，熔断器不应熔断；在出现短路时，应立即熔断；在电流发生正常变动（如电动机起动过程）时，熔断器不应熔断；在用电设备持续过载时，应延时熔断。对熔断器的选用主要包括类型选择和熔体额定电流的确定。

选择熔断器的类型时，主要依据负载的保护特性和短路电流的大小。例如，用于保护照明和电动机的熔断器，一般是考虑它们的过载保护，这时，希望熔断器的熔化系数适当小些。

所以容量较小的照明线路和电动机宜采用熔体为铅锌合金的 RC1A 系列熔断器，而大容量的照明线路和电动机，除过载保护外，还应考虑短路时分断短路电流的能力。若短路电流较小时，可采用熔体为锡质的 RC1A 系列或熔体为锌质的 RM10 系列熔断器。用于车间低压供电线路的保护熔断器，一般是考虑短路时的分断能力。当短路电流较大时，宜采用具有高分断能力的 RL1 系列熔断器。当短路电流相当大时，宜采用有限流作用的 RT0 系列熔断器。

熔断器的额定电压要大于或等于电路的额定电压。

熔断器的额定电流要依据负载情况而选择。

①电阻性负载或照明电路，这类负载起动过程很短，运行电流较平稳，一般按负载额定电流的 1~1.1 倍选用熔体的额定电流，进而选定熔断器的额定电流。

②电动机等感性负载，这类负载的起动电流为额定电流的 4~7 倍，一般选择熔体的额定电流为电动机额定电流的 1.5~2.5 倍。这样一般来说，熔断器难以起到过载保护作用，而只能用作短路保护，过载保护应用热继电器才行。

③为防止发生越级熔断，上、下级（供电干、支线）熔断器间应有良好的协调配合，为此，应使上一级（供电干线）熔断器的熔体额定电流比下一级（供电支线）大 1~2 个级差。

(4) 接触器。接触器是电气控制系统中使用量大、涉及面广的一种低压控制电器，用来频繁地接通和分断交直流主回路和大容量控制电路。主要控制对象是电动机，能实现远距离控制，并具有欠（零）电压保护。

1) 结构和工作原理。接触器主要由电磁系统、触头系统和灭弧装置组成，外形和结构简图如图 2-19 和图 2-20 所示。

图 2-19 交流接触器的外形

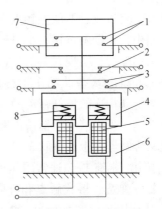

图 2-20 接触器的结构示意图
1—主触点；2—常闭辅助触点；3—常开辅助触点；4—动铁心；
5—电磁线圈；6—静铁心；7—灭弧装置；8—弹簧

工作原理：接触器根据电磁工作原理，当电磁线圈通电后，线圈电流产生磁场，使静铁心产生电磁吸力吸引衔铁，并带动触头动作，使常闭触头断开，常开触头闭合，两者是联动的。当电磁线圈断电时，电磁力消失，衔铁在释放弹簧的作用下释放，使触头复原，即常开触头断开，常闭触头闭合。接触器的图形符号、文字符号如图 2-21 所示。

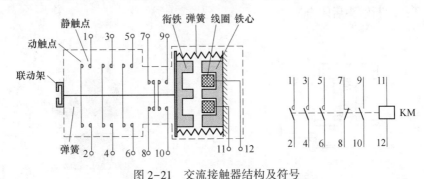

图 2-21 交流接触器结构及符号
1~6—内部为三组常开主触点；7，8—内部为常闭辅助触点；
9，10—内部为常开辅助触点；11，12—内部为控制线圈

2）交流接触器。接触器按其主触头所控制主电路电流的种类可分为交流接触器和直流接触器。

其中，交流接触器线圈通以交流电，主触头接通、分断交流主电路。直流接触器线圈通以直流电流，主触头接通、切断直流主电路。

由于交流接触器应用广泛，我们只介绍交流接触器。

交流接触器结构及符号如图 2-21 所示。在图 2-21 中，1~6 为主触点，7、8 为常闭辅助触点，9、10 为常开辅助触点，11、12 为线圈。图的右侧为其在图样中的符号。

交流接触器工作原理：

交流接触器主要由主触点、辅助触点和控制线圈组成，当给控制线圈通电时，线圈产

生磁场，磁场通过铁心吸引衔铁，而衔铁则通过连杆带动所有的动触点动作，与各自的静触点接触或断开。交流接触器的主触点允许流过的电流较辅助触点大，故主触点通常接在大电流的主电路中，辅助触点接在小电流的控制电路中。交流接触器的工作原理是利用电磁力与弹簧弹力相配合，实现触头的接通和分断的。交流接触器有两种工作状态：失电状态（释放状态）和得电状态（动作状态）。当吸引线圈通电后，使静铁心产生电磁吸力，衔铁被吸合，与衔铁相连的连杆带动触头动作，使常闭触头断接触器处于得电状态；当吸引线圈断电时，电磁吸力消失，衔铁再复开，使常开触头闭合，在弹簧作用下释放，所有触头随之复位，接触器处于失电状态。

（5）控制按钮。控制按钮通常用作短时接通或断开小电流控制电路的开关。控制按钮是由按钮帽、复位弹簧、桥式触点和外壳等组成。通常制成具有常开触点和常闭触点的复合式结构，其结构示意图如图2-22所示。指示灯式按钮内可装入信号灯显示信号；紧急式按钮装有蘑菇形钮帽，以便于紧急操作。旋钮式按钮是用手扭动旋钮来进行操作的。

按钮帽有多种颜色，一般红色用作停止按钮，绿色用作启动按钮。按钮主要根据触点数、使用场合及颜色来进行选择。

其外观如图2-23所示。按钮是一种常用的控制电器元件，常用来接通或断开"控制电路"（其中电流很小），从而达到控制电动机或其他电气设备运行目的的一种开关。其触点的动作规律是：当按下时，其动断触头先断开，动合触头后闭合；当松手时，则动合触头先断开，动断触头后闭合。

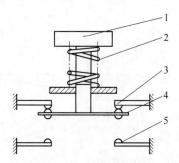

图2-22 控制按钮结构示意图
1—按钮帽；2—复位弹簧；3—常闭触头；
4—动触头；5—常开触头

图2-23 按钮

按钮开关的图形符号及文字符号如图2-24所示。

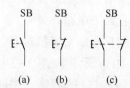

图2-24 图形符号及文字符号
(a) 常开触头；(b) 常闭触头；(c) 复式触头

（6）热继电器。热继电器是利用电流的热效应原理工作的，用于电动机或其他电气设备的过载保护和缺相保护的保护电器。但是由于热继电器中的发热元件有热惯性，在电路中不能做瞬时过载保护，更不能做短路保护，因此它不同于电流继电器和熔断器。

1）热继电器的结构。热继电器的结构如图2-25所示。

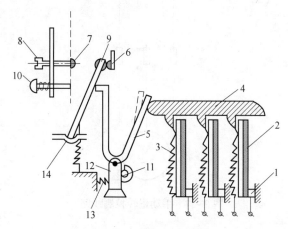

图2-25 热继电器的结构

1—接线端子；2—双金属片；3—热元件；4—导板；5—补偿双金属片；6—常闭静触头；7—常开触头；8—复位螺钉；9—常闭动触头；10—复位按钮；11—调节旋钮；12—支撑件；13—压簧；14—推杆

2）工作原理。过载电流通过热元件后，使双金属片加热弯曲去推动动作机构来带动触点动作，从而将电动机控制电路断开实现电动机断电停车，起到过载保护的作用。鉴于双金属片受热弯曲过程中，热量的传递需要较长的时间，因此，热继电器不能用作短路保护，而只能用作过载保护。热继电器的符号如图2-26所示。

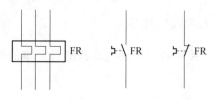

图2-26 热继电器的符号

2.2.4.2 三相异步电动机手动正转控制线路

手动正转控制线路只能控制电动机单向启动和停止。并带动生产机械的运动朝一个方向旋转或运动。

对于小容量的电动机，只要电动机接上额定电压就可以直接起动。这种起动方式，称为全压起动。对于三相鼠笼式异步电动机，全压起动时流过电动机的电流，将远远超过电动机的额定电流，大约为额定电流的5~7倍。过大的起动电流会引起线路上很大的电压降，要影响其他用电设备的正常运行。所以全压起动仅适用于容量较小的电动机。而对于容量较大的电动机应采用减压起动的方法。

手动正转控制线路如图 2-27 所示。

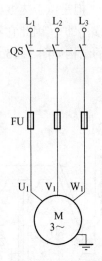

图 2-27　手动正转控制线路

(1) 线路的特点。

优点：简单、造价也低。

缺点：不能远距离操作且操作时不安全，易发生灼伤手的事故。

(2) 各低压电器的作用。手动开关 QS 起接通、断开电源用；熔断器作短路保护用。

(3) 线路的工作原理。

启动：合上刀开关 QS→电动机 M 接通电源启动运转。

停止：断开刀开关 QS→电动机 M 脱离电源停止运转。

2.2.4.3　三相异步电动机点动控制线路

点动控制线路是用按钮、接触器来控制电动机的最简单的控制线路。原理图如图2-28所示。

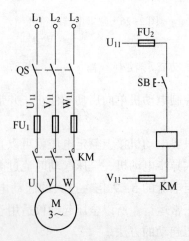

图 2-28　点动控制线路原理图

点动控制是指：按下按钮，电动机得电运转；松开按钮，电动机断电停转。这种控制方法常用于电动机的起重电机和车床拖板箱快速移动的电机控制。

控制线路通常是采用国家标准规定的电气图形符号和文字符号，画成控制线路原理图来表示。它是依据实物接线电路绘制的，用来表达控制线路的工作原理。上述的点动控制原理图可分成主电路和控制电路两大部分。主电路是从电源 L_1、L_2、L_3 经电源开关 QS、熔断器 FU_1、接触器 KM 的主触点到电动机 M 的电路，它流过的电路较大。由熔断器 FU_2、按钮 SB 和接触器 KM 的线圈组成控制电路，流过的电流较小。当电动机需点动时，先合上电源开关 QS，按下点动按钮 SB，接触器线圈 KM 便通电，衔铁吸合，带动它的三对常开主触点 KM 闭合，电动机 M 便接通电源起动运转。SB 按钮放开后，接触器线圈断电，衔铁受弹簧力的作用而复位，带动它的三对常开主触点断开，电动机断电停转。

点动控制线路的工作原理：合上电源开关 QS 后。

起动：按下 SB→KM 线圈通电→KM 主触点闭合→电动机 M 运转

停止：松开 SB→KM 线圈断电→KM 主触点断开→电动机 M 停转

2.2.4.4 三相异步电动机自锁控制线路

为了实现电动机的连续运行，可采用接触器自锁的正转控制线路，需要用接触器的一个常开辅助触点并联在起动按钮 SB_2 的两端，在控制电路中在串联一个停止按钮 SB_1，可以将电动机停止。电动机自锁控制线路如图 2-29 所示。

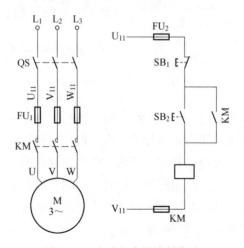

图 2-29 电动机自锁控制线路

这种接触器自锁的正转控制线路不但能使电动机连续运转，还具有欠压保护和失压（零压）保护的功能。

（1）欠压保护。"欠压"是指线路电压低于电动机应加的额定电压。"欠压保护"是指线路电压低于某一数值时，电动机能自动脱离电源电压停转，避免电动机在欠压下运行的一种保护。电动机为什么要有欠压保护呢？因为电动机运行时当电源电压下降，电动机的电流就会上升，电压下降越严重电流上升的也越严重，严重时会烧坏电动机。在接触器自锁的正转控制线路,当电动机运转时,电源电压降低到较低(一般在工作电压的85%以下)时,接触器线圈的磁通变得很弱,电磁吸力不足,动铁心在反作用弹簧的作用下释放,自锁触点断开,失去自锁,同时主触点也断开,电动机停转,得到了保护。

(2)失压(或零压)保护。"零压保护"是指电动机运行时,由于外界某种原因使电源临时停电时,能自动切断电动机电源。在恢复供电时,而不能让电动机自行起动,如果未加防范措施很容易造成人身事故。采用接触器自锁的正转控制线路,由于自锁触点和主触点在停电时已一起断开,控制电路和主电路都不会自行接通,所以在恢复供电时,如果没有按下按钮,电动机就不会自行起动。

(3)三相异步电动机自锁控制线路原理分析。合上 QS。

起动:

按下 SB_2 ⟶ KM线圈得电 ┬⟶ KM常开触头闭合自锁
　　　　　　　　　　　　　└⟶ KM主触头闭合 ⟶ 电动机M起动运行

停止:

按下 SB_1 ⟶ KM线圈失电 ┬⟶ KM自锁触头断开
　　　　　　　　　　　　　└⟶ KM主触头断开 ⟶ 电动机M停止运行

2.2.4.5　具有过载保护的自锁正转控制线路

过载保护是指当电动机出现过载时能自动切断电动机电源,使电动机停转的一种保护。最常用的是利用热继电器进行过载保护。电动机在运行过程中,如长期负载过大、操作频繁或断相运行等都可能使电动机定子绕组的电流超过它的额定值,但电流又未达到使熔断器熔断,将引起电动机定子绕组过热温度升高。如果温度超过允许温升,就会使绝缘损坏,电动机的使用寿命大为缩短,严重时甚至会烧坏电动机。

因此,对电动机必须采取过载保护的措施。

电动机在运行过程中,由于过载或其他原因使电流超过额定值,经过一定时间,串接在主电路中的热继电器 FR 的热元件受热发生弯曲,通过动作机构使串接在控制电路中的 FR 常闭触点断开,切断控制电路,接触器 KM 的线圈断电,主触点断开,电动机 M 便停转,达到了过载保护之目的。

具有过载保护的自锁正转控制线路如图 2-30 所示。

2.2.5　材料清单

材料清单见表 2-5。

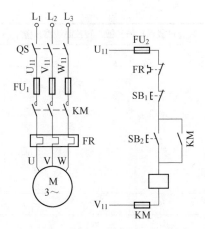

图 2-30 具有过载保护的自锁正转控制线路

表 2-5 材料清单

序号	名　　称	数量
1	XK-SX2C 型高级维修电工实训台	1 台
2	维修电工实训组件（XKDT11）	1 个
3	维修电工实训组件（XKDT12A）	1 个
4	三相异步电动机	1 台
5	跨接线	若干
6	万用表	1 块
7	接触器	若干
8	热继电器	若干
9	按钮	若干

2.2.6 任务实施

任务实施见表 2-6。

表 2-6 任务实施

具有过载保护的自锁正转控制线路任务书					
1. 根据要求完成所需器件清点并完成下表					
序号	器件名称	器件数量	器件参数	器件测量	备注

续表 2-6

具有过载保护的自锁正转控制线路任务书
2. 工作原理描述
3. 电路接线图的设计绘制
4. 操作过程记录
5. 通电试车过程记录
6. 任务小结

2.2.7 任务评价

任务评价见表 2-7。

表 2-7 任务评价

具有过载保护的自锁正转控制线路安装与调试评分表

器件清点和测量（10 分）

序号	重点检查内容	评分标准	分值	得分	备注
1	器件清点	器件清点错误一项扣 1 分	5		
2	器件测量	器件测量错误一项扣 1 分	5		
		小　计			

电路图设计（10 分）

序号	重点检查内容	评分标准	分值	得分	备注
1	主电路设计	主电路错误一处扣 1 分	5		
2	控制电路设计	控制电路设计错误一处扣 1 分	5		
		小　计			

工作原理描述（10 分）

序号	重点检查内容	评分标准	分值	得分	备注
1	主电路工作原理	是否完整酌情	5		
2	控制电路工作原理	是否完整酌情	5		
		小　计			

线路施工（30 分）

序号	重点检查内容	评分标准	分值	得分	备注
1	导线处理是否正确	错误一处扣 1 分	10		
2	导线安装是否牢固	错误一处扣 1 分	10		
3	线路是否正确	错误一处扣 1 分	10		
		小　计			

通电试车（30 分）

序号	重点检查内容	评分标准	分值	得分	备注
1	自锁功能	功能是否实现	10		
2	过载保护功能	功能是否实现	10		
3	停止功能	功能是否实现	10		
		小　计			

续表 2-7

具有过载保护的自锁正转控制线路安装与调试评分表

职业素养（10分）

序号	重点检查内容	评分标准	扣分	得分	备注
1	带电操作	一次扣5分扣完为止			
2	规范操作	一次扣1分扣完为止			
3	团队合作	酌情			
4	工位整洁	酌情			
总　计					

2.2.8 故障的排除

电动机控制线路故障排除时，要合理利用工具，例如万用表，可以使用万用表的交流电压档 AC 500V 测量电源是否正常，也可使用万用表的欧姆档测量线路、低压电器是否导通（注意：此时应断电）。

电动机的起动控制线路的安装与调试过程中存在的故障主要有：

故障1，在点动控制线路中，合上开关 QS，按下按钮 SB，接触器 KM 无动作，电动机不运行。

故障排除：断开电源，用万用表欧姆档，测量电路中5个熔断器是否正常，若导通，则逐个测量导线、QS 是否导通，若仍正常则，将万用表打到交流电压档 AC 500V，测量电源电压是否正常。

故障2，在自锁控制线路中，合上开关 QS，按一下按钮 SB_2，接触器 KM 闭合，电动机运行，松开 SB_2，KM 失电断开，电动机停止运行。

故障排除：故障是接触器 KM 没有自锁，导致电动机不能连续运行。此时，应使用万用表欧姆档检查与 SB_2 并联的接触器常开触点的上进、下出线路接线是否正常。

故障3，在点动控制线路中，按一下按钮 SB，接触器吸合，电动机得电起动运行，松开按钮，接触器 KM 主触头仍导通，电动机仍然运行。

故障排除：此故障是接触器主触头老化生锈，动作不灵，接触器线圈失电后，主触头不能分开，导致电动机仍然得电运行。维修方法是将接触器拆开，利用锉打磨接触器的主触点和辅助触点。

在进行线路安装时注意以下事项，可以减少故障的发生：

（1）接线时注意走线集中、减少架空和交叉，做到横平、竖直、转弯成直角。

（2）操作时不许用手触及各电气元件的导电部分及电动机的转动部分，以免触电及意外损伤。

（3）只有在断电的情况下，才可用万用表 Ω 档来检查线路的接线正确与否。

（4）螺旋式熔断器的接线要正确，以确保用电安全。

（5）训练应在规定的时间内完成，同时要做到安全操作和文明生产。

故障排除记录单见表 2-8。

表 2-8 故障排除记录单

	故障排除记录单	
故障 1	故障现象	
	故障原因	
	故障的排除过程	
故障 2	故障现象	
	故障原因	
	故障的排除过程	
故障 3	故障现象	
	故障原因	
	故障的排除过程	
故障 4	故障现象	
	故障原因	
	故障的排除过程	

2.2.9 任务拓展

连续运行和点动控制混合的控制线路：

机床设备在正常工作时，一般需要电动机处在连续运转状态，但在试车或调整刀具与工件相对位置时，要求电动机能点动控制。实现这种工艺要求的线路是连续与点动混合控制的正转控制线路。它们的主电路相同。

图 2-31 为连续与点动控制线路。

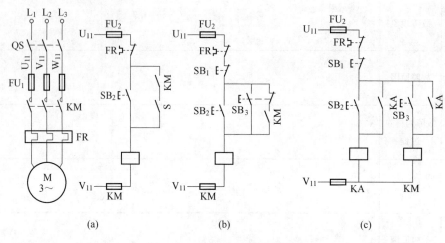

图 2-31　连续与点动控制线路

图 2-31（a）是在接触器正转控制线路的基础上，把一个手动开关 S 串联在自锁电路中实现的。

工作原理：先合上电源开关 QS，当 S 断开时，按下 SB_2，为点动控制；当 S 合上时，按一下 SB_2，为具有自锁的连续控制。

图 2-31（b）是在具有自锁的控制电路中增加一个复合按钮 SB_3。

需要点动运行时，按下 SB_3 点动按钮，其常闭触点先断开自锁电路，常开触发后闭合接通起动控制电路，KM 接触器线圈得电，主触点闭合，接通三相电源，电动机起动运转。当松开点动按钮 SB_3 时，KM 线圈失电，KM 主触点断开，电动机停止运转。

若需要电动机连续运转，由停止按钮 SB_1 及起动按钮 SB_2 控制，接触器 KM 的辅助触点起自锁作用。

图 2-31（c）是在控制电路中增加了一个点动按钮 SB_3 和一个中间继电器 KA。

先合上电源开关 QS，连续控制为：

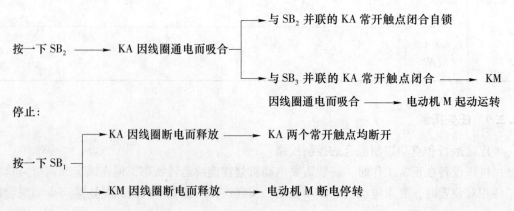

点动控制为：

起动：按下 SB_3 ——→ KM 因线圈通电而吸合 ——→ 电动机 M 起动运转

停止：空开 SB_3 ——→ KM 因线圈断电而释放 ——→ 电动机 M 断电运转

以上 3 种控制线路各有优缺点。

图 2-31（a）比较简单，由于连续与点动都是用同一按钮 SB_2 控制的，所以如果疏忽了开关 S 的操作，就会引起混淆。

图 2-31（b）虽然将连续与点动按钮分开了，但是当接触器贴心因剩磁而发生缓慢释放时，就会使点动控制变成连续控制。例如，在松开 SB_3 时，它的常闭触点应该是在 KM 自锁触点断开后才闭合，如果接触器发生缓慢释放，KM 自锁触点还未断开，SB_3 的常闭触点已经闭合，KM 线圈就不再断电而变成连续控制了。在某些极限状态下这是十分危险的。所以这种控制线路虽然简单却并不可靠。

图 2-31（c）多用了一个中间继电器 KA 相比之下虽不够经济，然而可靠性却大大提高了。

2.2.10 任务小结

本任务学习了低压电器和电动机的起动控制线路。

低压电器各式各样、品牌众多，常用的有断路器、控制按钮、接触器、热继电器、熔断器等。

电动机接通电源后由静止状态逐渐加速到稳定状态的过程称为电动机的起动。

直接起动又叫全压起动，它是通过开关或接触器将额定电压直接加在电动机的定子绕组上而使电动机起动的方法。

电动机的起动控制线路主要有手动正转控制线路、点动控制线路、自锁控制线路、过载保护的控制起动线路和既能连续运行又能点动控制的控制线路。

任务 2.3　三相异步电动机两地控制线路安装调试

2.3.1　任务描述

一些大型生产设备为了保证操作安全，常要求能多个地点进行控制操作，如锅炉房的鼓风机、循环水泵电动机、机床等均需在现场就地控制和在控制室远程控制。本任务通过接触器、按钮实现电动机的两地控制。

2.3.2　任务目标

（1）认识交流接触器主触头与辅助触头的连接方法及所起到的作用。
（2）理解两地控制的含义、作用以及实现的方法。
（3）学会两地控制中起动按钮与停止按钮的连接方式以及注意事项。

2.3.3　任务分析

以两地控制为例。若一台电动机要实现甲、乙两地控制，其中 SB_{11}、SB_{12} 为安装在甲地的起动按钮和停止按钮，SB_{21}、SB_{22} 为安装在乙地的起动按钮和停止按钮。线路的特点是两地的起动按钮 SB_{11}、SB_{21} 并联在一起；停止按钮 SB_{12}、SB_{22} 串联在一起。这样就可以在甲、乙两地起动、停止同一台电动机，达到操作方便的目的。

2.3.4 任务相关知识

三相异步电动机甲、乙两地控制电路工作原理：按下甲地起动按钮 SB_{11} 或者乙地起动按钮 SB_{21}，接触器 KM 线圈得电，接触器 KM 主触点闭合，接触器 KM 辅助常开触点闭合，起自锁作用，电动机保持运行；按下甲地停止按钮 SB_{12} 或者乙地停止按钮 SB_{22}，接触器 KM 线圈失电，接触器 KM 主触点断开，接触器 KM 辅助常开触点断开，取消自锁作用，电动机停止运行。

图 2-32 是三相异步电动机甲、乙两地控制电路电路图。电路中将甲、乙两地起动按钮 SB_{11}、SB_{21} 并联连接，停止按钮 SB_{12}、SB_{22} 串联连接，因此甲、乙两地都可以分别起动和停止同一台三相异步电动机。

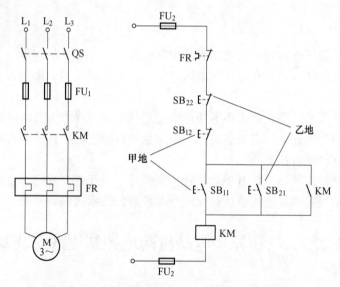

图 2-32　三相异步电动机甲、乙两地控制电路图
QS—断路器；FR—热继电器；FU—熔断器；KM—接触器；
SB_{11}—甲地起动按钮；SB_{12}—甲地停止按钮；SB_{21}—乙地起动按钮；SB_{22}—乙地停止按钮

2.3.5 材料清单

材料清单见表 2-9。

表 2-9　材料清单

序号	名　　称	数量
1	XK-SX2C 型高级维修电工实训台	1 台
2	维修电工实训组件（XKDT11）	1 个
3	维修电工实训组件（XKDT12A）	1 个

续表 2-9

序号	名　称	数量
4	三相异步电动机	1 台
5	跨接线	若干

2.3.6 任务实施

任务实施见表 2-10。

表 2-10　任务实施

三相异步电动机两地控制线路安装调试任务书

1. 根据要求完成所需器件清点并完成下表

序号	器件名称	器件数量	器件参数	器件测量	备注

2. 工作原理描述

3. 电路接线图的设计绘制

续表 2-10

三相异步电动机两地控制线路安装调试任务书
4. 操作过程记录
5. 通电试车过程记录
6. 任务小结

2.3.7 任务评价

任务评价见表 2-11。

表 2-11 任务评价

三相异步电动机两地控制线路安装调试评分表					
器件清点和测量（10 分）					
序号	重点检查内容	评分标准	分值	得分	备注
1	器件清点	器件清点错误一项扣 1 分	5		
2	器件测量	器件测量错误一项扣 1 分	5		
		小　计			
电路图设计（10 分）					
序号	重点检查内容	评分标准	分值	得分	备注
1	主电路设计	主电路错误一处扣 1 分	5		
2	控制电路设计	控制电路设计错误一处扣 1 分	5		
		小　计			

续表 2-11

三相异步电动机两地控制线路安装调试评分表

工作原理描述（10 分）

序号	重点检查内容	评分标准	分值	得分	备注
1	主电路工作原理	是否完整酌情	5		
2	控制电路工作原理	是否完整酌情	5		
	小　　计				

线路施工（30 分）

序号	重点检查内容	评分标准	分值	得分	备注
1	导线处理是否正确	错误一处扣 1 分	10		
2	导线安装是否牢固	错误一处扣 1 分	10		
3	线路是否正确	错误一处扣 1 分	10		
	小　　计				

通电试车（30 分）

序号	重点检查内容	评分标准	分值	得分	备注
1	甲地功能	功能是否实现	15		
2	乙地功能	功能是否实现	15		
	小　　计				

职业素养（10 分）

序号	重点检查内容	评分标准	扣分	得分	备注
1	带电操作	一次扣 5 分扣完为止			
2	规范操作	一次扣 1 分扣完为止			
3	团队合作	酌情			
4	工位整洁	酌情			
	总　　计				

2.3.8 故障的排除

2.3.8.1 常见故障

（1）按下 SB_{11}、SB_{21} 电动机不能起动。

（2）电动机只能点动控制。

（3）按下 SB_{11} 电动机不能起动，按下 SB_{21} 电动机能起动。

2.3.8.2 故障检修流程图

故障检修流程图如图 2-33 所示。

故障排除记录单见表 2-12。

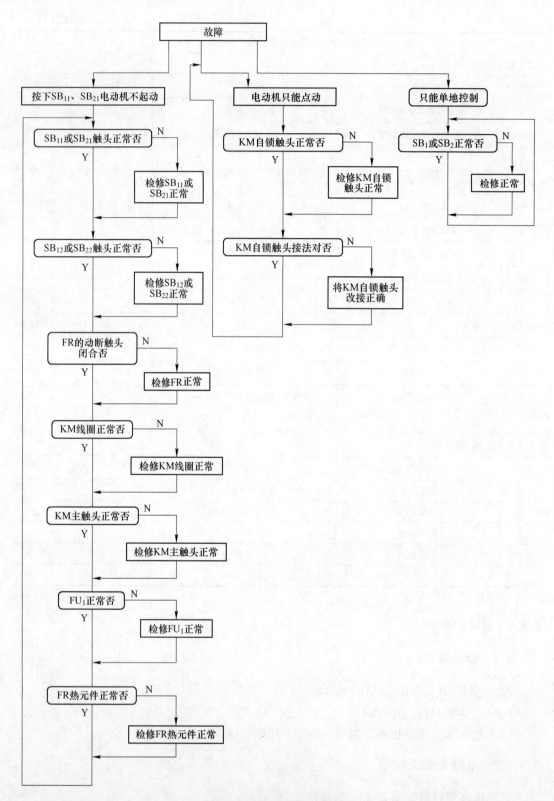

图 2-33 故障检修流程图

表 2-12 故障排除记录单

	故障排除记录单	
故障 1	故障现象	
	故障原因	
	故障的排除过程	
故障 2	故障现象	
	故障原因	
	故障的排除过程	
故障 3	故障现象	
	故障原因	
	故障的排除过程	
故障 4	故障现象	
	故障原因	
	故障的排除过程	

2.3.9 任务拓展

三相异步电动机的三地控制电路实现。在原有两地控制电路基础上，再加入一对按钮，设计一个"三相异步电动机三地控制电路"。

完成任务：

(1) 完成电路图的设计；
(2) 完成"三相异步电动机三地控制电路"原理讲解。

2.3.10 任务小结

电动机要实现两地控制，将两地起动按钮 SB_{11}、SB_{21} 并联连接，停止按钮 SB_{12}、SB_{22} 串联连接，因此两地都可以分别起动和停止同一台三相异步电动机。

任务实施过程中应先断开电源，按图进行实验接线；先通电调试控制回路，看接触器的动作是否正确；控制回路调试正确无误后，再通电调试主回路，看电动机的运转是否正确。实施过程中不能带电进行接线操作，应先调试控制回路后调试主回路。

项目 3　三相异步电动机的复杂控制技能训练

任务 3.1　三相异步电动机正反转控制线路安装调试

3.1.1　任务描述

生产机械往往要求运动部件可以实现正反两个方向的运动,这就要求拖动电动机能做正、反向旋转,如伸缩门、升降机等。由电机原理可知,改变电动机三相电源的相序,就能改变电动机的转向。本任务通过接触器、按钮实现电动机的正反转控制。

3.1.2　任务目标

(1) 认识交流接触器与辅助触头的连接方法及所起到的作用。
(2) 理解互锁的含义、作用以及实现互锁的方法。
(3) 学会实现电动机正反转的各种方法以及注意事项。

3.1.3　任务分析

电动机要实现正反转控制,将其电源的相序中任意两相对调即可(我们称为换相),通常是 V 相不变,将 U 相与 W 相对调,为了保证两个接触器动作时能够可靠调换电动机的相序,接线时应使接触器的上口接线保持一致,在接触器的下口调相。由于将两相相序对调,故须确保两个 KM 线圈不能同时得电,否则会发生严重的相间短路故障,因此必须采取联锁。为安全起见,常采用按钮联锁(机械)与接触器联锁(电气)的双重联锁正反转控制线路;使用了按钮联锁,即使同时按下正反转按钮,调相用的两接触器也不可能同时得电,机械上避免了相间短路。另外,由于应用的接触器联锁,所以只要其中一个接触器得电,其长闭触点就不会闭合,这样在机械、电气双重联锁的应用下,电机的供电系统不可能相间短路,有效地保护了电动机,同时也避免在调相时相间短路造成事故,烧坏接触器。

本任务分为:
(1) 接触器互锁的正反转控制电路安装调试;
(2) 按钮联锁控制电动机正反转控制电路安装调试;
(3) 接触器按钮双重互锁正反转控制电路安装调试;
(4) 接触器按钮双重互锁正反转控制电路安装调试。

3.1.4　任务相关知识

(1) 三相电动机正反转工作原理:当电动机的三相定子绕组通入三相对称交流电后,将产生一个旋转磁场,该旋转磁场切割转子绕组,从而在转子绕组中产生感应电流(转子绕组是闭合通路），载流的转子导体在定子旋转磁场作用下将产生电磁力，从而在电动机转轴上形成电磁转矩，驱动电动机旋转，并且电动机旋转方向与旋转磁场方向相同。将其电源的相序中任意两相对调，产生的旋转磁场会改变方向，从而电动机的旋转方向也会改变方向，从而实现电动机的正反转。

（2）互锁：三相电动机正反转控制中同一个电动机的"正转"和"反转"两个按钮应实现互锁控制，即按下其中一个按钮时，另一个按钮必须自动断开电路，这样可以有效防止两个按钮同时通电造成机械故障或人身伤害事故。互锁在电动机上的应用是接触器互锁正反转电路，双重互锁正反转电路，按钮互锁正反转电路。

3.1.4.1 接触器互锁的正反转控制电路安装调试

图 3-1 是接触器互锁的正反转控制电路。用正向接触器 KM_1 和反向接触器 KM_2 来完成主回路两相电源的对调工作，从而实现正反转的转换。

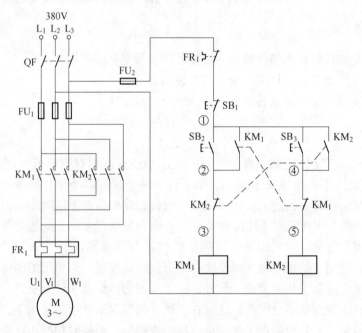

图 3-1 接触器互锁的正反转控制电路

QF—断路器；FR—热继电器；FU—熔断器；SB_1—停止按钮；SB_2—正转按钮；SB_3—反转按钮；KM_1—正转接触器；KM_2—反转接触器

在控制回路中，利用正向接触器 KM_1 的常闭触点 KM_1（④-⑤）控制反向接触器 KM_2 的线圈，利用反向接触器 KM_2 的常闭触点 KM_2（②-③）控制正向接触器 KM_1 的线圈，从而达到相互锁定的作用。这两对常闭触点称为互锁触点，这两个常闭触点组成的电路称为互锁环节。

当电源开关闭合后，按下正向起动按钮 SB_2，正向接触器 KM_1 线圈通电吸合，主回路常开主触点闭合，电动机正向起动运行。同时，控制回路的常开辅助触点 KM_1（①-②）

闭合实现自锁；常闭辅助触点 KM_1（④-⑤）断开，切断反向接触器 KM_2 线圈电路，实现互锁。

当需要停车时，按下停止按钮 SB_1，切断正向接触器 KM_1 线圈电源，接触器 KM_1 衔铁释放，常开主触点恢复断开状态，电动机停止运转，同时自锁触点也恢复断开状态，自锁作用解除，为下一次起动做好准备。

反向起动的过程只需按下反向起动按钮 SB_3 即可完成反向起动的全过程，其步骤与正向起动相似。

互锁触点的作用：假设在按下正向起动按钮 SB_2，电动机正向起动后，由于某种原因（如误操作），又把反向起动按钮 SB_3 也按下时，由于正向接触器的互锁触点 KM_1（④-⑤）已断开，反向接触器不会接通。显然，如果没有互锁辅助触点 KM_1（④-⑤）的互锁作用，反向接触器 KM_2 线圈就会通电，那就必然造成主回路正、反向接触器的6个常开触点全部闭合，发生电源短路事故，这是绝对不允许的！同理，反向起动后，反向接触器 KM_2 的常闭辅助触点就切断了正向接触器 KM_1 的线圈回路，可以有效地防止正向接触器错误地接通主回路而发生电源短路事故。

这种控制线路的缺点是操作不方便：在改变电动机转向时，需要先按停止按钮，然后再按起动按钮，才能使电动机改变转向。

3.1.4.2 按钮联锁控制电动机正反转控制电路安装调试

图 3-2 是按钮联锁控制电动机正反转控制电路图，需要采用复合式按钮。复合按钮的动作特点是先断后通，即动断触点先断开，动合触点再闭合。正向复合按钮 SB_2 的常闭触点串接在反向接触器 KM_2 的线圈回路中，而反向复合按钮 SB_3 的常闭触点串接在正向接触器 KM_1 的线圈回路中。这样，在按下 SB_2 时，只有正向接触器 KM_1 的线圈可以得电吸合，而按下 SB_3 时，只有反向接触器 KM_2 可以得电吸合。

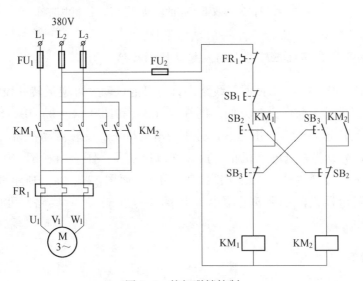

图 3-2 按钮联锁控制

如果发生误操作,比如,同时按下两个起动按钮 SB_2 和 SB_3,则两个接触器都不会得电吸合,可以防止发生两个接触器同时吸合而引起的主回路短路事故。

这种控制线路的优点是操作方便,当需要改变电动机转向时,不必再先按停止按钮了。但是这种线路也容易发生短路故障,例如当接触器 KM_1 主触点因故延迟释放或不能释放时,如果此时按反转按钮 SB_3,使接触器 KM_2 线圈通电,其主触点闭合,就会发生正反转接触器同时吸合,造成两相电源短路。可见,这种线路也不够安全。

3.1.4.3 接触器按钮双重互锁正反转控制电路安装调试

把图 3-1 和图 3-2 结合起来,就变成具有接触器按钮双重互锁的正反转控制线路,如图 3-3 所示。该线路把上述两个线路的优点结合起来,既可不按停止按钮而直接按反转按钮进行反向起动,当正转接触器发生熔焊故障时又不会发生相间短路故障。

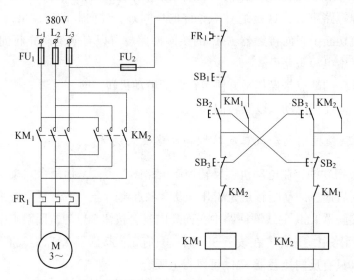

图 3-3 按钮接触器双重互锁的电动机正反转控制电路图

3.1.4.4 三相异步电动机正反转点动、起动控制线路

图 3-4(a)和图 3-4(b)所示线路,具有可逆点动、可逆运转并设有按钮及接触器触点双重互锁机构,操作比较方便。图 3-4(a)和图 3-4(b)中,SB_1、SB_2 分别为电动机正、反转起动按钮,SB_3、SB_4 分别为电动机正、反转点动按钮;SB_5 为停止按钮;KM_1、KM_2 分别为控制电动机正、反转交流接触器。图 3-4(b)中,按钮至接触器与熔断器之间,控制线只需 4 根(图中①~④),而图 3-4(a)所示线路,需 7 根。如果按钮到受控电器之间的距离很长,采用图 3-4(b)所示线路,可以节省较多的导线。

3.1.5 材料清单

材料清单见表 3-1。

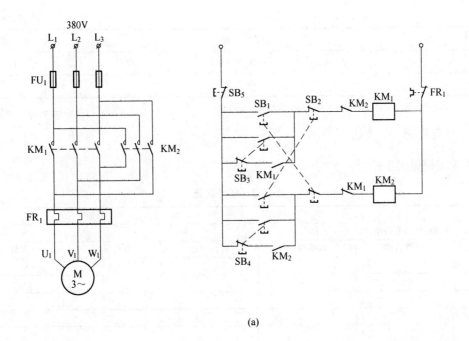

(a)

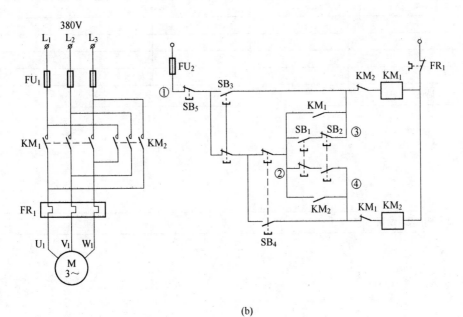

(b)

图 3-4 正反转点动、起动控制线路

表 3-1 材料清单

序号	名　　称	数量
1	XK-SX2C 型高级维修电工实训台	1 台
2	维修电工实训组件（XKDT11）	1 个
3	维修电工实训组件（XKDT12A）	1 个

续表 3-1

序号	名　称	数量
4	三相异步电动机	1台
5	跨接线	若干

3.1.6　任务实施

任务实施见表 3-2。

表 3-2　任务实施

（　　　　　　　　　　　　　　）任务书					
1. 根据要求完成所需器件清点并完成下表					
序号	器件名称	器件数量	器件参数	器件测量	备注
2. 工作原理描述					
3. 电路接线图的设计绘制					

续表 3-2

( ) 任务书
4. 操作过程记录
5. 通电试车过程记录
6. 任务小结

3.1.7 任务评价

任务评价见表 3-3。

表 3-3 任务评价

( ) 评分表					
器件清点和测量（10 分）					
序号	重点检查内容	评分标准	分值	得分	备注
1	器件清点	器件清点错误一项扣 1 分	5		
2	器件测量	器件测量错误一项扣 1 分	5		
小　计					
电路图设计（10 分）					
序号	重点检查内容	评分标准	分值	得分	备注
1	主电路设计	主电路错误一处扣 1 分	5		
2	控制电路设计	控制电路设计错误一处扣 1 分	5		
小　计					
工作原理描述（10 分）					
序号	重点检查内容	评分标准	分值	得分	备注
1	主电路工作原理	是否完整酌情	5		
2	控制电路工作原理	是否完整酌情	5		
小　计					

续表 3-3

（   ）评分表

线路施工（30 分）

序号	重点检查内容	评分标准	分值	得分	备注
1	导线处理是否正确	错误一处扣 1 分	10		
2	导线安装是否牢固	错误一处扣 1 分	10		
3	线路是否正确	错误一处扣 1 分	10		
	小　　计				

通电试车（30 分）

序号	重点检查内容	评分标准	分值	得分	备注
1	正转功能	功能是否实现	10		
2	反转功能	功能是否实现	10		
3	互锁功能	功能是否实现	10		
	小　　计				

职业素养（10 分）

序号	重点检查内容	评分标准	扣分	得分	备注
1	带电操作	一次扣 5 分扣完为止			
2	规范操作	一次扣 1 分扣完为止			
3	团队合作	酌情			
4	工位整洁	酌情			
	总　　计				

3.1.8　故障的排除

常规排除步骤：

（1）对电路控制板接入三相电源并接通电动机；

（2）检测接线无误，闭合开关对连接好的实物板进行通电检测；操作正转起动停止、反转起动，观察电动机的运行情况；

（3）若出现故障，必须断电检测，对出现的故障分析原因并让学生重新检测、调试，再通电，直到试车成功；

（4）最后总评故障现象及其原因。

出现的问题及解决办法：

（1）控制电路无自锁。这是因为交流接触器 KM_1（或 KM_2）的常开没有与开关 SB_2（SB_3）并联，并与线圈串联在一起，当出现此问题时，检测是 KM_1 无自锁还是 KM_2 无自锁，若是 KM_1 则应检测 KM_1 的常开，否则查看 KM_2。

（2）控制电路无互锁。这是因为两个交流接触器 KM_1、KM_2 的常闭没有互相控制彼此的线圈电路，即 KM_1（KM_2）的常闭没有串联于 KM_2（KM_1）的线圈电路中。

（3）控制电路不带电。可能为控制电路没有取相（回相）造成的，此时可以检查一下控制电路，当按下开关 SB_2 或 SB_3 时，是否通路，若通路，则检测熔断器是否正常。

(4）主电路不带电。此时可能开关没有闭合，或熔断器已烧坏，也有可能是主触点接触不良，可用万用表测量，然后确定问题所在。

(5）电路少相。表现为电动机转速慢，并产生较大的噪声，此时可以此测量三相电路，确定少相的线路，并加以调整。

(6）电路短路。此问题最为严重，必须对整个电路进行测量检查。

常见故障举例见表3-4。

表3-4 常见故障举例

故障现象	故障原因
电动机正反转均缺相，KM_1、KM_2 线圈吸合均正常	KM_1、KM_2 主电路回路中的共用回路有线路、触点、熔丝或电动机绕组损坏
电动机正转缺相，反转正常，KM_1、KM_2 线圈吸合均正常	KM_1 主电路回路中 U 相的线路或触点损坏
电动机正转正常，反转缺相，KM_1、KM_2 线圈吸合均正常	KM_2 的主电路回路中 V 相的线路或触点损坏
电动机正反转均无，KM_1、KM_2 线圈均不吸合	KM_1、KM_2 线圈不得电，控制电路回路中的共用回路有线路、触点或熔丝损坏
电动机正反转正常，KM_1、KM_2 线圈均吸合但电动机无法停止	控制回路中的 SB_3 触点或 SB_3 触点上下线路短路
电动机无正转，反转正常，KM_1 线圈不吸合，KM_2 线圈吸合正常	电动机正转控制回路中的线路、触点或线圈损坏
电动机正转正常，反转点动，KM_1 线圈吸合正常，KM_2 线圈吸合点动	KM_2 控制回路无法自锁，自锁回路中有线路或触点损坏
电动机正转正常，无反转，KM_1 线圈吸合正常，KM_2 线圈不吸合	电动机反转控制回路中的线路、触点或线圈损坏

故障排除记录单见表3-5。

表3-5 故障排除记录单

故障排除记录单		
故障1	故障现象	
	故障原因	
	故障的排除过程	

续表 3-5

故障排除记录单

故障 2	故障现象	
	故障原因	
	故障的排除过程	
故障 3	故障现象	
	故障原因	
	故障的排除过程	
故障 4	故障现象	
	故障原因	
	故障的排除过程	

3.1.9 任务拓展

工作台自动往返控制线路。

行程开关是位置开关的一种，其作用与按钮相同，能将机械信号转换成电信号，只是触点的动作不靠手动操作，而是用产生机械运动部件的碰撞，使触点动作来实现接通和分断电路，达到一定的控制目的。通常被用来限制机械运动的位置和行程，使运动机械按一定位置或行程自动停止，反向运动，变速运动或者自动往返运动等。

在原有正反转电路基础上，加入行程开关 SQ_1、SQ_2 设计一个"工作台自动往返控制线路"。

设计要求：

（1）实现电动机的起动、停止及正反转功能；

（2）电动机正转过程中触发行程开关 SQ_1，电动机反转；

（3）电动机反转过程中触发行程开关 SQ_2，电动机正转。

完成任务：
(1) 完成电路图的设计；
(2) 完成"工作台自动往返控制线路"原理讲解。

3.1.10 任务小结

电动机要实现正反转控制，将其电源的相序中任意两相对调即可（我们称为换相），通常是 V 相不变，将 U 相与 W 相对调，为了保证两个接触器动作时能够可靠调换电动机的相序，接线时应使接触器的上口接线保持一致，在接触器的下口调相。由于将两相相序对调，故须确保两个 KM 线圈不能同时得电，否则会发生严重的相间短路故障，因此必须采取互锁。而联锁的方式有接触器互锁、按钮互锁、按钮接触器双重联锁等方式。

任务实施过程中应先断开电源，按图进行实验接线；先通电调试控制回路，看接触器的动作是否正确；控制回路调试正确无误后，再通电调试主回路，看电动机的运转是否正确。实施过程中不能带电进行接线操作，应先调试控制回路后调试主回路。

任务 3.2 三相异步电动机降压起动线路安装调试

3.2.1 任务描述

三相异步电动机直接起动的起动电流大，对供电变压器影响较大，容量较大的三相鼠笼异步电动机一般都采用降压起动。降压起动就是将电源电压适当降低后，再加到电动机的定子绕组上进行起动，待电动机起动结束或将要结束时，再使电动机的电压恢复到额定值。

降压起动的目的是为了减少起动电流，但电动机的起动转矩也将降低。因此，降压起动仅适用于空载或轻载下的起动。降压起动常用方法有：定子绕组串电阻（或电抗器）降压起动、Y-△降压起动、自耦变压器降压起动和延边三角形降压起动等。本任务主要介绍通过接触器、时间继电器、按钮实现电动机的串电阻和Y-△降压起动。

3.2.2 任务目标

(1) 了解电动机的Y接法和△接法。
(2) 理解串电阻降压起动、Y-△降压起动的含义、作用以及方法。
(3) 学会实现降压起动的各种方法以及注意事项。

3.2.3 任务分析

三相异步电动机采用全压起动（直接起动），控制电路简单；但当电动机容量较大时，直接起动导致起动电流很大，很可能会使电网电压降低而影响其他电器的正常运行，因此一般不允许采用全压起动，而应采用降压起动。降压起动是指利用起动设备将电压适当降低后加到电动机的定子绕组上进行起动，待电动机起动运转后，再使其电压恢复到额定值正常运转，由于电流随电压的降低而减小，因此降压起动达到了减小起动电流的目的。

本任务分为：

(1) 三相异步电动机定子串电阻降压起动手动控制线路安装调试；
(2) 三相异步电动机定子串电阻降压起动自动控制线路安装调试；
(3) 时间继电器控制三相异步电动机Y-△起动线路安装调试；
(4) 接触器控制三相异步电动机Y-△起动线路安装调试。

3.2.4 任务相关知识

3.2.4.1 三相异步电动机定子串电阻降压起动手动控制线路

电动机起动时在三相定子电路中串入电阻，使电动机定子绕组电压降低，限制了起动电流，待电动机转速上升到一定值时，将电阻切除，使电动机在额定电压下稳定运行。

图3-5是定子串电阻降压起动手动控制线路，它的工作过程如下：按起动按钮 SB_1，接触器 KM_1 的线圈通电，接触器 KM_1 的自锁触点和主触点闭合，电动机串电阻起动。在接触器 KM_1 的线圈通电的同时，经过一定的时间，电动机起动结束或将要结束时，按下 SB_2，接触器 KM_2 的线圈通电，接触器 KM_2 的主触点闭合，并实现自锁，将串接电阻切除，电动机接入正常电压，并进入正常稳定运行。

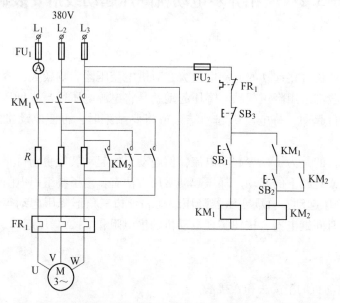

图3-5 三相异步电动机定子串电阻降压起动手动控制线路

定子串电阻降压起动虽然降低了起动电流，但起动转矩也降低了，这种起动方法只适用于空载或轻载起动。

3.2.4.2 三相异步电动机定子串电阻降压起动自动控制线路

如图3-6所示，合上电源开关 QF，按下起动按钮 SB_1，接触器 KM_1 与时间继电器 KT 的线圈同时通电，KM_1 主触头闭合，由于 KM_2 线圈有回路中串有时间继电器 KT 延时闭合的动合触头而不能吸合，这时电动机定子绕组中串有电阻 R，进行降压起动，电动机的转速逐步升高，当时间继电器 KT 达到预先整定的时间后，其延时闭合的动合触头闭合，

KM₂吸合，主触头闭合，将起动电阻 R 短接，电动机便处在额定电压下全压运转，通常 KT 的延时时间为 4~8s。

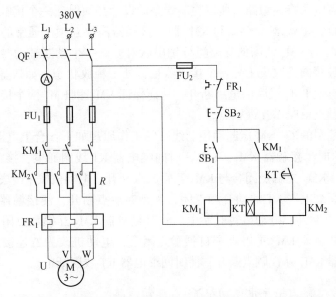

图 3-6 三相异步电动机定子串电阻降压起动自动控制线路

3.2.4.3 时间继电器控制三相异步电动机丫-△起动线路

额定运行为三角形接法且容量较大的电动机可以采用丫-△降压起动。电动机起动时，定子绕组按丫连接，每相绕组的电压降为三角形连接时的 $1/\sqrt{3}$，待转速升高到一定值时，改为△连接，直到稳定运行。丫-△降压起动控制线路如图 3-7 所示。

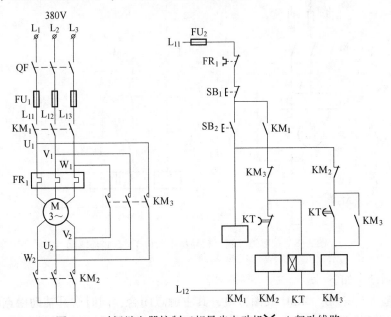

图 3-7 时间继电器控制三相异步电动机丫-△起动线路

从图 3-7 中可以看到主电路中有三组主触点，其中接触器 KM_2 和 KM_3 主触点一定不能同时闭合，因为开关 QF 合上电源，接触器 KM_1 主触点闭合后，接触器 KM_2 和 KM_3 如同时闭合，意味着电源将被短路。所以控制线路的设计必须保证一个接触器吸合时，另一个接触器不能吸合，也就是说 KM_2 和 KM_3 两个接触器需要互锁。通常的方法是在控制线路中，接触器 KM_2 与 KM_3 线圈的支路里分别串联对方的一个动断辅助触点。这样，每个接触器线圈能否被接通，取决于另一个接触器是否处于释放状态，如接触器 KM_2 已接通，它的动断辅助触点把 KM_3 线圈的电路断开，从而保证 KM_2 和 KM_3 两个接触器不会同时吸合，这一对动断触点就叫作互锁触点。

时间继电器控制的Y-△降压起动控制线路的工作原理如下：合上电源开关 QF，按下起动按钮 SB_2，这时，接触器 KM_1、KM_2、时间继电器 KT 线圈通电，接触器 KM_1 主触点和自锁触点闭合。KM_2 主触点闭合与 KM_3 互锁触点断开，电动机按Y接法起动，经过整定延时时间后，时间继电器 KT 的动合触点闭合和动断触点断开，使接触器 KM_2 线圈断电，接触器 KM_2 主触点断开，电动机暂时断电，同时接触器 KM_2 互锁触点闭合，使得接触器 KM_3 线圈通电，接触器 KM_3 主触点和自锁触点闭合，电动机改为△连接，然后进入稳定运行，同时接触器 KM_3 互锁触点断开，使时间继电器 KT 线圈断电。

3.2.4.4 接触器控制三相异步电动机Y-△起动线路

接触器控制三相异步电动机Y-△起动线路如图 3-8 所示。

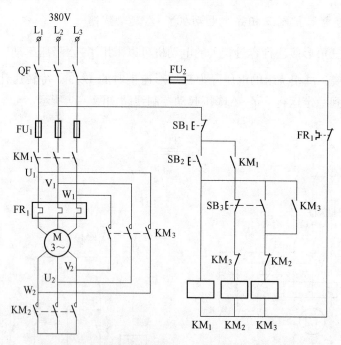

图 3-8 接触器控制三相异步电动机Y-△起动线路

线路控制动作如下：

Y接法起动：按 SB_2，KM_1 线圈得电，其主触点闭合，同时常开辅助触点闭合，实现自锁。在 KM_1 线圈得电的同时，KM_2 的主触点闭合，同时 KM_2 的常闭辅助触点断开，断

开了 KM_3 线圈的通路,实现了联锁。由于 KM_1、KM_2 主触点同时闭合,实现了电动机的Y接法起动。

△接法运行:经过一定的时间,电动机起动结束或将要结束时,按一下 SB_3,SB_3 常闭触点分断,KM_2 线圈失电,KM_2 主触点复位,解除电动机绕组Y。同时 KM_2 常闭辅助触点复位,解除联锁。按 SB_3 的同时,SB_3 常开触点闭合,KM_3 线圈得电,其主触点闭合,电动机绕组△接法运行。同时,KM_3 常闭触点分断,断开了 KM_2 的线圈通路,实现了联锁。

停车:按 SB_1,控制电路断电,各接触器同时释放,电动机停车。

Y接法的起动电流仅为△接法的 1/3,从而限制了起动电流,但是Y接法的起动转矩为△接法的 1/3,所以Y-△起动只适用空载或轻载起动。

3.2.5 材料清单

材料清单见表 3-6。

表 3-6 材料清单

序号	名 称	数量
1	XK-SX2C 型高级维修电工实训台	1 台
2	维修电工实训组件(XKDT11)	1 个
3	维修电工实训组件(XKDT12A)	1 个
4	三相异步电动机	1 台
5	跨接线	若干

3.2.6 任务实施

任务实施见表 3-7。

表 3-7 任务实施

三相异步电动机降压起动线路安装调试任务书					
1. 根据要求完成所需器件清点并完成下表					
序号	器件名称	器件数量	器件参数	器件测量	备注

续表 3-7

三相异步电动机降压起动线路安装调试任务书
2. 工作原理描述
3. 电路接线图的设计绘制
4. 操作过程记录
5. 通电试车过程记录

续表 3-7

三相异步电动机降压起动线路安装调试任务书
6. 任务小结

3.2.7 任务评价

任务评价见表 3-8。

表 3-8 任务评价

三相异步电动机 Y-△ 降压起动线路安装调试评分表

器件清点和测量（10 分）

序号	重点检查内容	评分标准	分值	得分	备注
1	器件清点	器件清点错误一项扣1分	5		
2	器件测量	器件测量错误一项扣1分	5		
		小　计			

电路图设计（10 分）

序号	重点检查内容	评分标准	分值	得分	备注
1	主电路设计	主电路错误一处扣1分	5		
2	控制电路设计	控制电路设计错误一处扣1分	5		
		小　计			

工作原理描述（10 分）

序号	重点检查内容	评分标准	分值	得分	备注
1	主电路工作原理	是否完整酌情	5		
2	控制电路工作原理	是否完整酌情	5		
		小　计			

线路施工（30 分）

序号	重点检查内容	评分标准	分值	得分	备注
1	导线处理是否正确	错误一处扣1分	10		
2	导线安装是否牢固	错误一处扣1分	10		
3	线路是否正确	错误一处扣1分	10		
		小　计			

续表 3-8

三相异步电动机 Y-△ 降压起动线路安装调试评分表

通电试车（30分）

序号	重点检查内容	评分标准	分值	得分	备注
1	Y 接起动功能	功能是否实现	10		
2	△ 接手动起动功能	功能是否实现	10		
3	△ 接自动起动功能	功能是否实现	10		
	小　　计				

职业素养（10分）

序号	重点检查内容	评分标准	扣分	得分	备注
1	带电操作	一次扣 5 分扣完为止			
2	规范操作	一次扣 1 分扣完为止			
3	团队合作	酌情			
4	工位整洁	酌情			
	总　　计				

3.2.8 故障的排除

（1）接触器控制三相异步电动机 Y-△ 起动线路 Y 起动过程正常，但按下 SB_3 后电动机发出异常声音转速也急剧下降，这是为什么？

分析现象：接触器切换动作正常，表明控制电路接线无误。问题出现在接上电动机后，从故障现象分析，很可能是电动机主回路接线有误，使电路由 Y 接法转到 △ 接法时，送入电动机的电源相序改变了，电动机由正常起动突然变成了反序电源制动，强大的反向制动电流造成了电动机转速急剧下降和异常声音。

处理故障：核查主回路接触器及电动机接线端子的接线顺序。

（2）线路空载试验工作正常，接上电动机试车时，一起动电动机，电动机就发出异常声音，转子左右颤动，立即按 SB_1 停止，停止时 KM_2 和 KM_3 的灭弧罩内有强烈的电弧现象。

分析现象：空载试验时接触器切换动作正常，表明控制电路接线无误。问题出现在接上电动机后，从故障现象分析是由于电动机缺相所引起的。电动机在 Y 起动时有一相绕组为接入电路，电动机造成单相起动，由于缺相绕组不能形成旋转磁场，使电动机转轴的转向不定而左右颤动。

处理故障：检查接触器接点闭合是否良好，接触器及电动机端子的接线是否紧固。

（3）时间继电器控制三相异步电动机 Y-△ 起动线路空载试验时，一按起动按钮 SB_2，KM_2 和 KM_3 就切换不能吸合。

分析现象：一起动 KM_2 和 KM_3 就反复切换动作，说明时间继电器没有延时动作，一按 SB_2 起动按钮，时间继电器线圈得电吸合，接点也立即动作，造成了 KM_2 和 KM_3 的相互切换，不能正常起动。

处理故障：问题出现在时间继电器的接点上。检查时间继电器的接线，发现时间继电器的接点使用错误，接到时间继电器的瞬动接点上了，所以一通电接点就动作，将线路改接到时间继电器的延时接点上故障排除。时间继电器往往有一对延时动作接点，还有一对瞬时动作接点，接线前应认真检查时间继电器的接点的使用要求。

故障排除记录单见表3-9。

表 3-9 故障排除记录单

故障排除记录单		
故障1	故障现象	
	故障原因	
	故障的排除过程	
故障2	故障现象	
	故障原因	
	故障的排除过程	
故障3	故障现象	
	故障原因	
	故障的排除过程	
故障4	故障现象	
	故障原因	
	故障的排除过程	

3.2.9 任务拓展

在本任务中，采用了三相异步电动机Y-△降压控制起动。在某些场合，可以通过三相异步电动机延边三角形联结实现降压起动，尝试完成该设计。

完成任务：

（1）完成电路图的设计；

（2）完成"三相异步电动机延边三角形降压起动控制线路"原理讲解。

3.2.10 任务小结

降压起动是指利用起动设备将电压适当降低后加到电动机的定子绕组上进行起动，待电动机起动运转后，再使其电压恢复到额定值正常运转，由于电流随电压的降低而减小，因此降压起动达到了减小起动电流的目的。但同时，由于电动机的转矩与电压的平方成正比，降压起动也将导致电动机的起动转矩大大降低。因此，降压起动需要在空载或轻载下进行。

任务实施过程中应先断开电源，按图进行实验接线；先通电调试控制回路，看接触器的动作是否正确；控制回路调试正确无误后，再通电调试主回路，看电动机的运转是否正确。实施过程中不能带电进行接线操作，应先调试控制回路后调试主回路。

项目 4　典型机床电气控制线路的故障检修

任务 4.1　CA6140 型车床电气控制线路故障检修

4.1.1　任务描述

机床也称为工作母机或工具机，机床的作用是制造机器，它们是主要应用于工业机械与设备、交通运输设备、初级金属制品和电气电子设备中机械制造的基础装备，广泛应用于各工业领域，其中金属切削机床按加工方式可分为车床、钻床、镗床、磨床等。本任务主要是掌握 CA6140 型车床电气控制线路故障排除方法。

4.1.2　任务目标

(1) 了解车床的功能和结构、运动形式。
(2) 熟悉 CA6140 型车床电气控制线路的组成，工作原理。
(3) 掌握 CA6140 型车床电气控制线路故障排除方法。

4.1.3　任务分析

车床是一种应用极为广泛的金属切削机床，主要用于加工各种回转表面和回转体的端面。如车削内外圆柱面、圆锥面、环槽及成形回转表面，车削端面及各种常用的螺纹，配有工艺装备还可加工各种特形面。其特点是车刀相对固定而工件高速旋转。

本任务分为：
(1) 了解车床主要结构、运动形式及电力拖动特点、控制要求。
(2) 识读 CA6140 型普通车床电气原理图。
(3) 进行 CA6140 型普通车床的电气控制线路安装与调试。
(4) 进行 CA6140 型普通车床的常见电气故障分析与检修。

4.1.4　任务相关知识

4.1.4.1　车床主要结构及运动形式

CA6140 型车床型号意义：

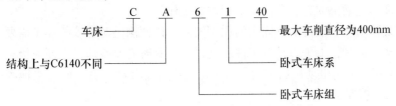

CA6140 型普通车床的外形及主要结构如图 4-1 所示。

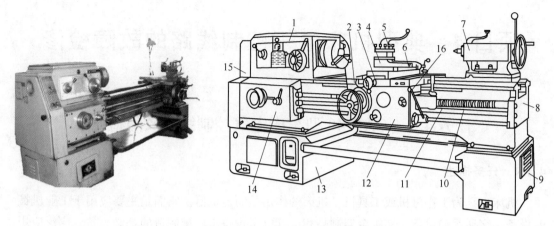

图 4-1　车床外形及主要结构

1—主轴箱；2—纵溜板；3—横溜板；4—转盘；5—方刀架；6—小溜板；7—尾架；8—床身；9—右床座；
10—光杠；11—丝杠；12—溜板箱；13—左床箱；14—进给箱；15—挂轮架；16—操纵手柄

CA6140 型普通车床主要由床身、主轴箱、进给箱、溜板箱、刀架、丝杠、光杠、尾架等部分组成。

车床的切削运动包括卡盘或顶尖带动工件旋转的主运动和溜板带动刀架及刀具的直线进给运动。车床工作时，绝大部分功率消耗在主轴运动上。车削速度是指工件与刀具接触点的相对速度；根据工件的材料性质、车刀材料及几何形状、工件直径、加工方式及冷却条件的不同，要求主轴有不同的切削速度。主轴变速是由主轴电动机经 V 带传递到主轴变速箱来实现的。CA6140 型车床的主轴正转速度有 24 种（10~1400r/min），反转速度有 12 种（14~1580r/min）。

车床的进给运动是刀架带动刀具的直线运动。溜板箱把丝杠或光杠的转动传递给刀架部分，变换溜板箱外的手柄位置，经刀架部分使车刀做纵向或横向进给。

车床的辅助运动为车床上除切削运动以外的其他一切必需的运动，如尾架的纵向移动、工件的夹紧与放松等。

4.1.4.2　电力拖动特点及控制要求

（1）主拖动电动机一般选用三相笼型异步电动机，不进行电气调速。

（2）采用齿轮箱进行机械有级调速。为了减小振动，主拖动电动机通过几条 V 带将动力传递到主轴箱。

（3）在车削螺纹时，要求主轴有正、反转，由主拖动电动机正反转或采用机械方法来实现。

（4）主拖动电动机的起动、停止采用按钮操作。

（5）刀架移动和主轴转动有固定的比例关系，以便满足对螺纹的加工需要。

（6）车削加工时，由于刀具及工件温度过高，有时需要冷却，因而应该配有冷却泵电动机，且要求在主拖动电动机启动后，方可决定冷却泵开动与否，而当主拖动电动机停止时，冷却泵应立即停止。

(7) 必须有过载、短路、欠压、失压保护。
(8) 具有安全的局部照明装置。

4.1.4.3 电气原理图分析

CA6140型普通车床电气原理图如图4-2所示。

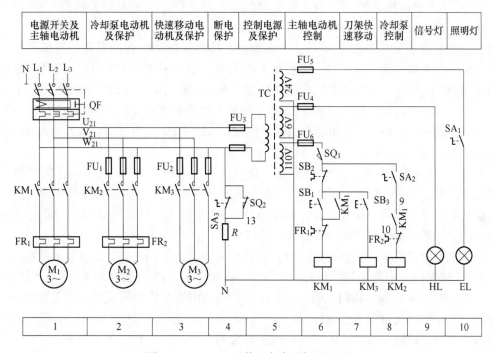

图4-2 CA6140型普通车床电气原理图

(1) 主电路分析。电源由漏电保护断路器QF引入。主轴电动机M_1的运转和停止由接触器KM_1的三个常开主触点的接通和断开来控制。电动机M_1的容量不大，所以采用直接起动。冷却泵电动机M_2的运转和停止由接触器KM_2的三个常开主触点来控制。快速移动电动机M_3的运转和停止由接触器KM_3的三个常开主触点来控制。漏电保护断路器QF的电源端应接熔断器作短路保护。冷却泵电动机M_2与快速移动电动机M_3的容量都很小，分别加装了熔断器FU_1与FU_2作短路保护。热继电器FR_1和FR_2分别作M_1与M_2的过载保护，其热元件接入各自的主电路中。快速移动电动机M_3是短时工作的，所以不需过载保护。

(2) 控制电路分析。控制电路采用110V交流电压供电，该电压是由380V电压经控制变压器TC降压而得。初级由熔断器FU_3、次级由FU_6作短路保护。先合上漏电保护断路器QF，设行程开关SQ_1的常开触点是闭合的。

1) 主轴电动机的控制。按下起动按钮SB_1，接触器KM_1线圈通电，KM_1的铁心吸合，主电路上KM_1的三个常开主触点闭合，主轴电动机M_1起动运转。同时，KM_1的一个常开辅助触点也闭合，进行自锁，保证主轴电动机M_1在松开起动按钮后能连续转动。按下停止按钮SB_2，接触器KM_1因线圈断电而释放，它的三个常开主触点断开，主轴电动机M_1便停止。热继电器FR_1的常闭触点串联在KM_1线圈的电路中，当主轴电动机M_1过载时，

FR_1 的常闭触点断开，KM_1 因线圈断电而释放，电动机 M_1 便停止。该电路有零压保护功能。在电源断电后，接触器 KM_1 释放，当电源电压再次恢复正常时，如不按下起动按钮 SB_1，则电动机不会自行起动，以免发生事故。该电路也有欠电压保护，当电源电压太低时，接触器 KM_1 因电磁吸力不足而自动释放，电动机 M_1 自行停止，以避免欠电压时电动机 M_1 因电流过大而烧坏。

2）冷却泵电动机的控制。当主轴电动机运转时，KM_1 的常开辅助触点（9-10）闭合，这时若需要冷却液，则可旋转转换开关 SA_2 使其闭合，则接触器 KM_2 线圈通电，铁心吸合，主电路上 KM_2 的三个常开主触点闭合，冷却泵电动机起动运转，给切削加工提供冷却液。当主轴电动机停车时，接触器 KM_1 释放，其常开触点（9-10）断开，冷却泵电动机 M_2 也同时停止。可见，只有当主轴电动机 M_1 起动后，冷却泵电动机 M_2 才能起动，两者之间存在联锁。热继电器 FR_2 的常闭触点串联在 KM_2 线圈的电路中，所以当冷却泵电动机过载时，FR_2 的常闭触点断开，接触器 KM_2 因线圈断电而释放，电动机 M_2 便停止，实现过载保护。接触器 KM_2 对冷却泵电动机也有欠压保护。

3）快速移动电动机的控制。快速移动是点动控制电路。按下按钮 SB_3，接触器 KM_3 线圈通电，铁心吸合，使主电路中 KM_3 的三个常开主触点闭合，快速移动电动机 M_3 运转，拖动刀架快速移动。松开按钮，KM_3 释放，M_3 即停止。快速移动的方向通过装在溜板箱上的十字形手柄扳到所需要的方向来控制。

除了主轴电动机 M_1 与冷却泵电动机 M_2 之间有上述的联锁之外，还有其他联锁。

4）联锁保护。钥匙式电源开关 SA_3 的触点（W_{21}-13）与行程开关 SQ_2 的常闭触点并联后与检漏电阻 R 串联。检漏电阻不通电流的条件是用钥匙将钥匙式电源开关旋转到 SA_3 断开位置，并且电气箱盖子已盖好，这时其盖子压下行程开关 SQ_2，常闭触点断开。只有在这种情况下，检漏电阻不通电，漏电保护开关 QF 才能合得上，以保证安全。SQ_1 为挂轮架安全行程开关。当装好挂轮架罩时，SQ_1 的常开触点闭合，控制电路才有电，电动机才可能起动。

（3）照明电路与信号指示电路分析。

照明电路采用 24V 交流电压。照明电路由开关 SA_1 接灯泡 EL 组成。灯泡 EL 的另一端必须接地，以防止照明变压器原绕组和副绕组间发生短路时可能发生的触电事故。熔断器 FU_5 是照明电路的短路保护。信号指示电路采用 6V 交流电压，指示灯泡 HL 接在控制变压器 TC 次级的 6V 线圈上，指示灯亮表示控制电路有电。熔断器 FU_4 是信号指示电路的短路保护。

4.1.4.4 CA6140型普通车床电气故障检修

（1）常见电气故障分析与检修。当需要打开配电盘壁龛门进行带电检修时，将 SQ_2 开关的传动杆拉出，断路器 QF 仍可合上。关上壁龛门后，SQ_2 复原恢复保护作用。

1）主轴电动机 M_1 不能起动，可按下列步骤检修：

检查接触器 KM_1 是否吸合，如果接触器 KM_1 吸合，则故障必然发生在电源电路和主电路上。可按下列步骤检修：

①合上断路器 QF，用万用表测量接触器受电端 U_{21}、V_{21}、W_{21} 点之间的电压，如果电压是 380V，则电源电路正常。当测量 U_{21} 与 W_{21} 之间无电压时，再测量 L_1、L_2、L_3 之间的

电压，若无，则说明电源故障，若有电压，则是断路器 QF 接触不良或连线断路。

修复措施：查明损坏原因，更换相同规格和型号断路器及连接导线。

②断开断路器 QF，用万用表电阻 R_{x1} 档测量接触器 KM_1 主触点间的阻值，如果阻值较小且相等，说明所测电路正常；否则，依次检查 FR_1、电动机 M_1 以及它们之间的连线。

修复措施：查明损坏原因，修复或更换同规格、同型号的热继电器 FR_1、电动机 M_1 及其之间的连接导线。

③检查接触器 KM_1 主触头是否良好，如果接触不良或烧损，则更换动、静触头或相同规格的接触器。

④检查电动机机械部分是否良好，如果电动机内部轴承等损坏，应更换轴承；如果外部机械有问题，可配合机修工进行维修。

若接触器 KM_1 不吸合，可按下列步骤检修：

首先检查 KM_3 是否吸合，若吸合说明 KM_1 和 KM_3 的公共控制电路部分正常，故障范围在 KM_1 的线圈部分支路；若 KM_3 也不吸合，就要检查照明灯和信号灯是否亮，若照明灯和信号灯亮，说明故障范围在控制电路上，若灯 HL、EL 都不亮，说明电源部分有故障，但不能排除控制电路也有故障。

2）主轴电动机 M_1 起动后不能自锁；当按下起动按钮 SB_1 时，主轴电动机能起动运转，但松开 SB_1 后，M_1 也随之停止。造成这种故障的原因是接触器 KM_1 的自锁触头接触不良或连接导线松脱。

3）主轴电动机 M_1 不能停车。造成这种故障的原因多是接触器 KM_1 的主触头熔焊；停止按钮 SB_2 击穿或线路中 SB_2 两端两点连接导线短路；接触器铁心表面粘牢污垢。可采用下列方法判明是哪种原因造成电动机 M_1 不能停车：若断开 QF，接触器 KM_1 释放，则说明故障为 SB_2 击穿或导线短接；若接触器过一段时间释放，则故障为铁心表面粘牢污垢；若断开 QF，接触器 KM_1 不释放，则故障为主触头熔焊。根据具体故障采取相应措施修复。

4）主轴电动机在运行中突然停车。这种故障的主要原因是由于热继电器 FR_1 动作。发生这种故障后，一定要找出热继电器 FR_1 动作的原因，排除后才能使其复位。引起热继电器 FR_1 动作的原因可能是：三相电源电压不平衡；电源电压较长时间过低；负载过重以及 M_1 的连接导线接触不良等。

5）刀架快速移动电动机不能起动。首先检查 FU_2 熔丝是否熔断；其次检查 KM_3 触头的接触是否良好；若无异常或按下 SB_3 时，接触器 KM_3 不吸合，则故障必定在控制电路中。这时依次检查点动按钮 SB_3 及接触器 KM_3 的线圈是否有断路现象即可。

（2）检修步骤及工艺要求。

1）在操作师傅的指导下对车床进行操作，了解车床的各种工作状态及操作方法。

2）在教师的指导下，参照电器位置图和机床接线图，熟悉车床电器元件的分布位置和走线情况。

3）在 CA6140 型车床电气控制线路上人为设置自然故障点。故障设置时应注意以下几点：

①人为设置的故障必须是模拟车床在使用中，由于受外界因素影响而造成的自然故障。

②切忌设置更改线路或更换电器元件等由于人为原因而造成的非自然故障。

③对于设置一个以上故障点的线路,故障现象尽可能不要相互掩盖。如果故障相互掩盖,按要求应有明显检查顺序。

④应尽量设置不容易造成人身或设备事故的故障点。

(3)注意事项。

1)熟悉 CA6140 型车床电气控制线路的基本环节及控制要求,认真观摩教师示范检修。

2)检修所用工具、仪表应符合使用要求。

3)排除故障时,必须修复故障点,但不得采用元件代换法。

4)检修时,严禁扩大故障范围或产生新的故障。

5)带电检修时,必须有指导教师监护,以确保安全。

4.1.5 材料清单

材料清单见表 4-1。

表 4-1 材料清单

序号	名　　称	数量
1	XK-SX2C 型高级维修电工实训台	1 台
2	维修电工实训组件(XKDT11)	1 个
3	维修电工实训组件(XKDT12A)	1 个
4	三相异步电动机	1 台
5	跨接线	若干

4.1.6 任务实施

CA6140 型普通车床电气线路安装、调试任务书见表 4-2。

表 4-2 CA6140 型普通车床电气线路安装、调试任务书

CA6140 型普通车床电气线路安装、调试任务书					
1. 根据要求完成所需器件清点并完成下表					
序号	器件名称	器件数量	器件参数	器件测量	备注

续表 4-2

CA6140 型普通车床电气线路安装、调试任务书

2. 工作原理描述

3. 电路接线图的设计绘制

4. 操作过程记录

5. 通电试车过程记录

续表 4-2

CA6140 型普通车床电气线路安装、调试任务书
6. 任务小结

4.1.7 任务评价

CA6140 型普通车床电气线路安装、调试评分表见表 4-3。

表 4-3　CA6140 型普通车床电气线路安装、调试评分表

CA6140 型普通车床电气线路安装、调试评分表					
器件清点和测量（10 分）					
序号	重点检查内容	评分标准	分值	得分	备注
1	器件清点	器件清点错误一项扣 1 分	5		
2	器件测量	器件测量错误一项扣 1 分	5		
小　计					
电路图设计（10 分）					
序号	重点检查内容	评分标准	分值	得分	备注
1	主电路设计	主电路错误一处扣 1 分	5		
2	控制电路设计	控制电路设计错误一处扣 1 分	5		
小　计					
工作原理描述（10 分）					
序号	重点检查内容	评分标准	分值	得分	备注
1	主电路工作原理	是否完整酌情	5		
2	控制电路工作原理	是否完整酌情	5		
小　计					
线路施工（30 分）					
序号	重点检查内容	评分标准	分值	得分	备注
1	导线处理是否正确	错误一处扣 1 分	10		
2	导线安装是否牢固	错误一处扣 1 分	10		
3	线路是否正确	错误一处扣 1 分	10		
小　计					

CA6140型普通车床电气线路安装、调试评分表

通电试车（30分）

序号	重点检查内容	评分标准	分值	得分	备注
1	主轴控制功能	功能是否实现	10		
2	冷却泵控制功能	功能是否实现	10		
3	快移控制功能	功能是否实现	5		
4	照明、信号指示控制	功能是否实现	5		
	小　　计				

职业素养（10分）

序号	重点检查内容	评分标准	扣分	得分	备注
1	带电操作	一次扣5分扣完为止			
2	规范操作	一次扣1分扣完为止			
3	团队合作	酌情			
4	工位整洁	酌情			
	总　　计				

4.1.8　故障的排除

故障排除记录单见表4-4。

表4-4　故障排除记录单

		故障排除记录单
故障1	故障现象	
	故障原因	
	故障的排除过程	
故障2	故障现象	
	故障原因	
	故障的排除过程	

续表 4-4

故障排除记录单

故障 3	故障现象	
	故障原因	
	故障的排除过程	
故障 4	故障现象	
	故障原因	
	故障的排除过程	

4.1.9 任务拓展

机床电气故障检修的一般方法如下：

(1) 检修前的故障调查。当工业机械发生电气故障后，切忌盲目随便动手检修。在检修前，通过问、看、听、摸来了解故障前后的操作情况和故障发生后出现的异常现象，以便根据故障现象判断出故障发生的部位，进而准确地排除故障。

问：询问操作者故障前后电路和设备的运行状况及故障发生后的症状；如故障是经常发生还是偶尔发生；是否有响声、冒烟、火花、异常振动等征兆；故障发生前有无切削力过大和频繁地起动、停止、制动等情况；有无经过保养检修或改动线路等。

看：查看故障发生前是否有明显的外观征兆，如各种信号；有指示装置的熔断器的情况；保护电器脱扣动作；接线脱落；触头烧蚀或熔焊；线圈过热烧毁等。

听：在线路还能运行和不扩大故障范围、不损坏设备的前提下，可通电试车，细听电动机、接触器和继电器等电器的声音是否正常。

摸：在刚切断电源后，尽快触摸检查电动机、变压器、电磁线圈及熔断器等，看是否有过热现象。

(2) 用逻辑分析法确定并缩小故障范围。检修简单的电气控制线路时，对每个电器元件、每根导线逐一进行检查，一般能很快找到故障点。但对复杂的线路而言，往往有上百个元器件，成千条连线，若采取逐一检查的方法，不仅需耗费大量的时间，而且也容易漏查。在这种情况下，可根据电路图，采用逻辑分析法，对故障现象具体分析，划出可疑范围，提高维修的针对性，就可以收到准而快的效果。分析电路时，通常先从主电路入手，了解工业机械各运动部件和机构采用了几台电动机拖动，与每台电动机相关的电器元器件有哪些，采用了何种控制，然后根据电动机主电路所用电器元器件的文字符号、图区号及控制要求，找到相应的控制电路。在此基础上，结合故障现象和线路工作原理，进行认真分析排查，即可迅速判定故障发生的可能范围。

当故障的可疑范围较大时，可在故障范围内的中间环节进行检查，来判断故障究竟是发生在哪一部分，从而缩小故障范围，提高检修速度。

（3）对故障范围进行外观检查。在确定了故障发生的可能范围后，可对范围内的电器元器件及连接导线进行外观检查，例如：熔断器的熔体熔断；导线接头松动或脱落；接触器和继电器的触头脱落或接触不良，线圈烧坏使表层绝缘纸烧焦变色，烧化的绝缘清漆流出；弹簧脱落或断裂；电气开关的动作机构受阻失灵等，都能明显地表明故障点所在。

（4）用试验法进一步缩小故障范围。经外观检查未发现故障点时，可根据故障现象，结合电路图分析故障原因，在不扩大故障范围、不损伤电气和机械设备的前提下，进行直接通电试验，或除去负载（从控制箱接线端子板上卸下）通电试验，以分清故障可能是在电气部分还是在机械等其他部分；是在电动机上还是在控制设备上；是在主电路上还是在控制电路上。一般情况下先检查控制电路，具体做法是：操作某一只按钮或开关时，线路中有关的接触器、继电器将按规定的动作顺序进行工作。若依次动作至某一电器元器件时，发现动作不符合要求，即说明该电器元器件或其相关电路有问题。再在此电路中进行逐项分析和检查，一般便可发现故障。待控制电路的故障排除恢复正常后，再接通主电路，检查控制电路对主电路的控制效果，观察主电路的工作情况有无异常等。

在通电试验时，必须注意人身和设备的安全。要遵守安全操作规程，不得随意触动带电部分，要尽可能切断电动机主电路电源，只在控制电路带电的情况下进行检查；如需电动机运转，则应使电动机在空载下运行，以避免工业机械的运动部分发生误动作和碰撞；要暂时隔断有故障的主电路，以免故障扩大，并预先充分估计到局部线路动作后可能发生的不良后果。

（5）用测量法确定故障点。测量法是维修电工工作中用来准确确定故障点的一种行之有效的检查方法。常用的测试工具和仪表有校验灯、测电笔、万用表、钳形电流表、绝缘电阻表等，主要通过对电路进行带电或断电时的有关参数（如电压、电阻、电流等）的测量，来判断电器元器件的好坏、设备的绝缘情况以及线路的通断情况。随着科学技术的发展，测量手段也在不断更新。例如，在晶闸管—电动机自动调速系统中，利用示波器来观察晶闸管整流装置的输出波形、触发电路的脉冲波形，就能很快判断系统的故障所在。

在用测量法检查故障点时，一定要保证各种测量工具和仪表完好，使用方法正确，还要注意防止感应电、回路电及其他并联支路的影响，以免产生误判断。

4.1.10 任务小结

以CA6140型普通车床电气线路的安装、调试及故障检修为例，了解机床电气故障检修的一般方法，在检修前通过故障调查——问、看、听、摸了解故障前后的操作情况和故障发生后出现的异常现象，用逻辑分析法确定并缩小故障范围，并通过正确使用各种测量工具和仪表测量故障点，进行故障检修。

任务 4.2　Z35 型摇臂钻床电气控制线路故障检修

4.2.1　任务描述

钻床是具有广泛用途的通用性机床，也是金属切削机床，它可对零件进行钻孔、扩孔、铰孔、锪平面和攻螺纹等加工。在钻床上配有工艺装备，还可以进行镗孔；配有万能工作台还能进行钻孔、扩孔、铰孔。本任务主要是掌握 Z35 型摇臂钻床控制线路故障排除方法。

4.2.2　任务目标

（1）了解钻床的功能和结构、运动形式。
（2）熟悉 Z35 型摇臂钻床控制线路的组成、工作原理。
（3）掌握 Z35 型摇臂钻床控制线路故障排除方法。

4.2.3　任务分析

钻床指主要用钻头在工件上加工孔的机床。钻床结构简单，加工精度相对较低，钻床的特点是工件固定不动，刀具做旋转运动。

本任务分为：
（1）了解 Z35 型摇臂钻床主要结构、运动形式及电力拖动特点、控制要求。
（2）识读 Z35 型摇臂钻床电气原理图。
（3）进行 Z35 型摇臂钻床的电气控制线路安装与调试。
（4）Z35 型摇臂钻床的常见电气故障分析与检修。

4.2.4　任务相关知识

4.2.4.1　摇臂钻床的主要结构和运动情况

Z35 型摇臂钻床型号的含义：

$$\underset{\text{钻床}}{Z}\ \underset{\text{摇臂钻床}}{3}\ \underset{\text{最大钻孔直径为50mm}}{5}$$

摇臂钻床的外形及主要结构如图 4-3 所示。在底座上固定有内立柱，内立柱的外面套有空心的外立柱。摇臂可以连同外立柱绕内立柱回转。摇臂与外立柱之间不能作相对转动。主轴箱可以在摇臂上沿导轨作水平移动。由于这些运动，可以方便地调整主轴上的钻头相对于工件的位置，以对准加工工件所需的加工孔中心。因此，一些大而重的多孔工件可以方便地在摇臂钻床上加工而无须移动工件。

工件不大时，可将其压紧在工作台上加工。如果工件较大，可以直接装在底座上加工。根据工件高度的不同，摇臂借助于丝杆可带着主轴箱沿外立柱上下升降。在升降之前，应自动将摇臂松开，再进行升降。当达到升降所需位置时，摇臂自动夹紧在立柱上。Z35 摇臂钻床摇臂升降时的松开与夹紧依靠机械机构自动进行。摇臂连同外立柱绕内立柱

 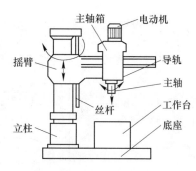

图 4-3 摇臂钻床的外形及主要结构

的回转运动依靠人力推动进行,但必须先将外立柱松开。主轴箱沿摇臂上导轨的水平移动也是手动的,也必须先将主轴箱松开。当需要进行加工时,应将外立柱夹紧在内立柱上,同时,应将主轴箱夹紧在摇臂上。这样,加工时主轴位置不会走动,刀具也不会振动。所以摇臂钻床的运动包括:主运动即主轴带动钻头的旋转运动;进给运动即钻头的上下移动;辅助运动有主轴箱沿摇臂水平移动、摇臂沿外立柱上下移动和摇臂连同外立柱一起相对于内立柱的回转。

4.2.4.2 电力拖动特点及控制要求

(1) 钻床的运动部件较多,采用多台电动机拖动。

(2) 钻床有时用来攻丝,所以要求主轴可以正反转。主轴的正反转一般通过正反转摩擦离合器来实现。Z35 摇臂钻床靠手柄推压弹簧和杠杆控制正反转摩擦离合器。因此,对于主轴电动机来说,只要求单方向旋转。主轴一般采用三相笼型异步电动机来拖动,用变速机构调节主轴转速和进刀量,主轴变速和进给变速的机构都在主轴箱内。

(3) 外立柱的松紧和主轴箱的松紧是依靠液压推动松紧机构同时进行的。

4.2.4.3 Z35 摇臂钻床电气原理图分析

Z35 摇臂钻床电气原理图如图 4-4 所示。

(1) 主电路分析。电源由转换开关 QS 引入,为机床加工做准备。整个机床由熔断器 FU_1 作短路保护。

从主电路中看到共有四台电动机。M_1 为冷却泵电动机,给加工工件提供冷却液,由转换开关 SA_2 直接控制。M_2 为主轴电动机,由接触器 KM_1 的常开主触点控制其起停,热继电器 FR 作过载保护。M_3 为摇臂升降电动机,由接触器 KM_2 和 KM_3 的常开主触点控制其正反转。M_4 为立柱放松与夹紧电动机,由接触器 KM_4 和 KM_5 的常开主触点控制其正反转。电动机 M_3 和 M_4 都是短时运行的,所以不加过载保护。M_3、M_4 及控制回路共用熔断器 FU_2 作短路保护。因为外立柱和摇臂要绕内立柱回转,所以除了冷却泵以外,其他的电源都通过汇流排 A 引入。

(2) 控制电路分析。为了安全起见,控制电路采用的电源电压为交流 127V,由变压器 TC 将 380V 的交流电压降压而得。

Z35 摇臂钻床控制电路中采用十字开关 SA_1 操作,它有控制集中的优点,还能实现主

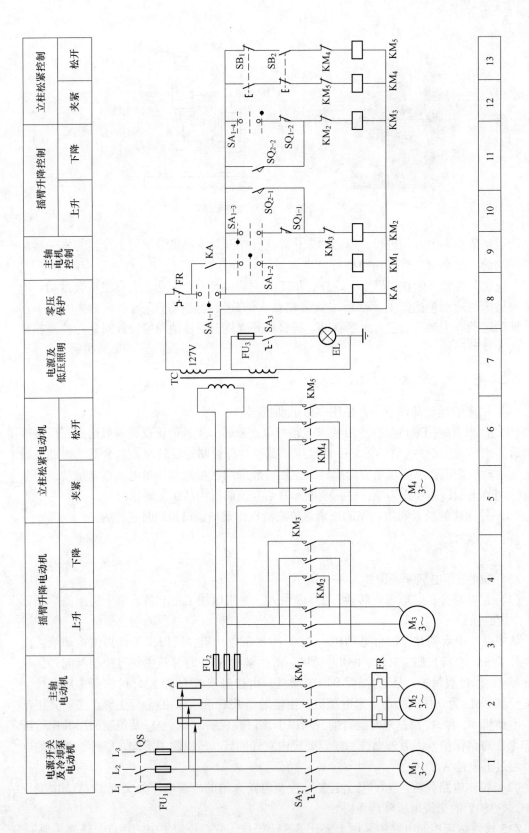

图 4-4 Z35摇臂钻床电气原理图

轴起停和摇臂升降等运动间的联锁,因为每次它只能扳到一个方向,接通一个方向的电路。十字开关由十字手柄和四个微动开关组成。十字手柄有五个位置,即上、下、左、右和中。各个位置的工作情况见表4-5。

表4-5 十字手柄各个位置的工作情况

手柄位置	实物位置	接通微动开关的触点	工作情况
中	✢	都不通	停止
左	✢	SA_{1-1}	零压保护
右	✢	SA_{1-2}	主轴运转
上	✢	SA_{1-3}	摇臂上升
下	✢	SA_{1-4}	摇臂下降

1) 零压保护。每次合电源或电源中断后又恢复时,必须将十字开关向左扳一次。这时,微动开关触点 SA_{1-1} 接通,零压继电器 KA 因线圈通电而吸合并自锁。当机床工作时,十字手柄不在左边位置。这时,若电源断电,零压继电器 KA 释放,其自锁触点也断开;当电源恢复时,继电器 KA 不会自行吸合,控制电路不会自行通电,这样便可防止可能发生的当电源中断后又恢复时,机床自行起动的危险。

2) 主轴电动机运转。将十字开关扳向右边,微动开关 SA_{1-2} 闭合,接触器 KM_1 因线圈通电而吸合,主轴电动机 M_2 起动后运转。主轴的正反转由主轴箱上的摩擦离合器手柄操作。摇臂钻床的钻头旋转和上下移动都由主轴电动机拖动。将十字开关扳到中间位置,SA_{1-2} 断开,主轴电动机 M_2 停止。

3) 摇臂的升降。钻头与工件之间的相对高度不合适时,可用摇臂升降来调整。欲使摇臂上升,可将十字开关扳向上边,微动开关触点 SA_{1-3} 闭合,接触器 KM_2 因线圈通电而吸合,电动机 M_3 正转,带动升降丝杆正转。升降丝杆与摇臂松紧的机构如图4-5所示。升降丝杆开始正转时,因升降螺母也跟着旋转,所以摇臂不会上升。下面的辅助螺母因不能旋转而向上移动,通过拨叉使传动松紧装置的轴逆时针方向转动,结果,松紧装置将摇臂松开。在辅助螺母向上移动时,带动传动条向上移动。当传动条压上升降螺母后,升降

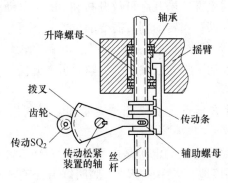

图4-5 摇臂升降前后的放松与加紧机构示意图

螺母就不能再转动了，而只能带动摇臂上升。在辅助螺母上升而转动拨叉时，拨叉又转动组合开关 SQ_2 的轴，使触点 SQ_{2-2} 闭合，为夹紧做准备。这时，KM_2 的常闭触点是断开的，所以接触器 KM_3 还不会吸合。

当摇臂上升到所需要的位置时，将十字开关扳回到中间位置，这时接触器 KM_2 因线圈断电而释放，其常闭触点闭合，又因触点 SQ_{2-2} 已闭合，接触器 KM_3 因线圈通电而吸合，电动机 M_3 反转使辅助螺母向下移动，一方面带动传动条下移而与升降螺母脱离接触，升降螺母又随丝杆空转，摇臂停止上升；另一方面辅助螺母下移时，通过拨叉使传动松紧装置的轴顺时针方向转动，结果，松紧装置将摇臂夹紧；同时，拨叉通过齿轮转动组合开关 SQ_2 的轴，使摇臂夹紧时触点 SQ_{2-2} 断开，接触器 KM_3 释放，电动机 M_3 停止。

欲使摇臂下降，可将十字开关扳向下边，微动开关触点 SA_{1-4} 闭合，接触器 KM_3 因线圈通电而吸合，电动机 M_3 反转，带动升降丝杆反转。开始时，升降螺母也跟着旋转，所以摇臂不会下降。下面的辅助螺母向下移动，通过拨叉使传动松紧装置的轴顺时针方向转动，结果，松紧装置也是先将摇臂松开。在辅助螺母向下移动时，带动传动条向下移动。当传动条压上升降螺母后，升降螺母也不转了，带动摇臂下降。辅助螺母下降而转动拨叉时，拨叉又转动组合开关 SQ_2 的轴，使触点 SQ_{2-1} 闭合，为夹紧做准备。这时，KM_3 的常闭触点是断开的。

当摇臂下降到所需要的位置时，将十字开关扳回到中间位置，这时触点 SA_{1-4} 断开，接触器 KM_3 因线圈断电而释放，其常闭触点闭合，又因触点 SQ_{2-1} 已闭合，接触器 KM_2 因线圈通电而吸合，电动机 M_3 正转使辅助螺母向上移动，带动传动条上移而与升降螺母脱离接触，升降螺母又随丝杆空转，摇臂停止下降；辅助螺母上移时，通过拨叉使传动松紧装置的轴逆时针方向转动，结果松紧装置将摇臂夹紧；同时，拨叉通过齿轮转动组合开关 SQ_2 的轴，使摇臂夹紧时触点 SQ_{2-1} 断开，接触器 KM_2 释放，电动机 M_3 停止。

限位开关 SQ_1 是用来限制摇臂升降的极限位置。当摇臂上升到极限位置，触点 SQ_{1-1} 断开，接触器 KM_2 因线圈断电而释放，电动机 M_3 停转，摇臂停止上升。当摇臂下降到极限位置，触点 SQ_{1-2} 断开，接触器 KM_3 因线圈断电而释放，电动机 M_3 停转，摇臂停止下降。

4）立柱的松开与夹紧。立柱的松开与夹紧是靠电动机 M_4 的正反转经过液压装置来完成的。当需要松开立柱时，可按下按钮 SB_1，接触器 KM_4 因线圈通电而吸合，电动机 M_4 正转，通过齿式离合器，M_4 带动齿轮式油泵旋转，从一定方向送出高压油，经一定的油路系统和传动机构将外立柱松开。松开后可放开按钮 SB_1，电动机停转，即可用人力推动摇臂连同外立柱绕内立柱转动。当转动到所需要的位置时，可按下 SB_2，接触器 KM_5 因线圈通电而吸合，电动机 M_4 反转，通过齿式离合器，M_4 带动齿轮式油泵反向旋转，从另一方向送出高压油，在液压推动下将立柱夹紧。夹紧后可放开按钮 SB_2，接触器 KM_5 因线圈断电而释放，电动机 M_4 停转。

Z35 摇臂钻床的主轴箱在摇臂上的松开与夹紧和立柱的松开与夹紧由同一台电动机（M_4）和同一液压机构进行。

(3) 照明电路分析。照明电路的电源是由变压器 TC 将 380V 的交流电压降至 36V 安全电压提供。照明灯 EL 端接地，以保证安全。照明灯由开关 SA_3 控制，由熔断器 FU_3 作短路保护。

4.2.4.4 Z35型摇臂钻床故障检修

(1) 常见电气故障分析与检修。

1) 主轴电动机不能起动。故障原因可能为：

①熔断器 FU_1 的熔体烧断，应更换熔体。

②微动开关 SA_{1-2} 损坏或接触不良，应修复或更换。

③零压继电器 KA 的触点接触不良或接线松脱，结果控制电路没有电压。

④电源电压太低。

⑤接触器 KM_1 的主触点接触不良或接线松脱。

2) 主轴电动机不能停止。一般是由于接触器的常开主触点熔焊造成，应更换接触器 KM_1 的主触点。

3) 摇臂升降以后不能完全夹紧。故障原因可能为：

①行程开关 SQ_2 动触点位置发生偏移。当摇臂升降完毕尚未完全夹紧时，触点 SQ_{2-1}（原摇臂下降）或触点 SQ_{2-2}（原摇臂上升）过早地断开，所以不能完全夹紧。将 SQ_2 的动触点 SQ_{2-1} 和 SQ_{2-2} 调到适当位置，故障便可排除。

②机床经检修后转动行程开关 SQ_2 的齿轮与拨叉上的扇形齿轮的啮合位置发生了偏移，当摇臂尚未完全夹紧时，触点 SQ_{2-1} 或触点 SQ_{2-2} 就过早地断开了，未到夹紧位置电动机 M_3 就停转了。

4) 摇臂升降方向与十字开关标志的扳动方向相反。这一故障的原因是升降电动机的电源相序接反了。发生这一故障是危险的，应立即断开电源开关。因为这时十字开关的触点和终端限位开关的触点都被行程开关 SQ_2 的触点短路，失去控制作用和终端保护作用。以十字开关扳到摇臂下降的位置为例，摇臂升降电动机 M_3 起动后摇臂不下降而往上升方向移动，这时 SQ_{2-2} 闭合，将十字开关扳回零位，触点 SQ_{1-4} 断开，接触器 KM_3 不会释放，摇臂还是继续上升，直到将上升终端限位开关撞开，SQ_{2-2} 仍不断开，摇臂仍然上升，应立即断开电源。

5) 摇臂升降不能停止。摇臂升降到所需的位置时，再将十字开关扳回中间位置，摇臂却继续升降，到终端限位开关触点断开也无济于事。这时，应及时断开电源，以免发生事故。这是因为检修时误将触点 SQ_{2-1} 和 SQ_{2-2} 的接线互换了。以十字开关扳到下降位置为例，KM_3 吸合，电动机 M_3 反转，摇臂先松开后下降，松开后应该是触点 SQ_{2-1} 闭合，为夹紧做准备，接线接错后变为 SQ_{2-2} 闭合，后将十字开关扳回中间位置及终端限位开关触点 SQ_{1-2} 断开也不会停。

6) 摇臂升降电动机正反转交替运转不停。当摇臂升降完毕以后，摇臂升降电动机 M_3 应反向旋转将摇臂夹紧，夹紧完毕 M_3 应停止。但是如果行程开关 SQ_2 的两个触点 SQ_{2-1} 和 SQ_{2-2} 调得太近，当上升（或下降）到所需位置时，将十字开关扳回零位，接触器 KM_2（下降为 KM_3）已释放，触点 SQ_{2-2}（下降为 SQ_{2-1}）已闭合，KM_3（下降为 KM_2）吸合，电动机反转（下降为正转）将摇臂夹紧，夹紧完毕 SQ_{2-2}（下降为 SQ_{2-1}）断开，KM_3（下降为 KM_2）释放。但由于电动机等的机械惯性，电动机及传动部分仍再转动一小段距离，使行程开关触点 SQ_{2-1}（下降为 SQ_{2-2}）因太近而被接通，接触器 KM_2（下降为 KM_3）又吸合，电动机又正转（下降为反转）起来，经过很短距离电动机 M_3 因触点 SQ_{2-1}（下降

为 SQ_{2-2}）断开而减速，由于机械惯性再转过一小段距离，使行程开关触点 SQ_{2-2}（下降为 SQ_{2-1}）因太近而被接通，接触器 KM_3（下降为 KM_2）又吸合，电动机又反转（下降为正转），接着循环下去，使夹紧与放松的动作重复不停。应仔细调整行程开关的两个触点 SQ_{2-1} 和 SQ_{2-2} 之间的距离，使它们不要太近，故障便可排除。

7）立柱松紧电动机不能起动。故障原因可能为：

①熔断器 FU_2 熔体已断，应更换熔体。

②按钮 SB_1 或 SB_2 的触点接触不良。

③接触器 KM_4 或 KM_5 的触点接触不良。

8）立柱松紧电动机不能停止。这一故障一般是因为接触器 KM_4 或 KM_5 的主触点熔焊，应立即断开电源，更换主触点。

(2) 检修步骤及工艺要求。

1）在操作师傅的指导下，对钻床进行操作，了解钻床的各种工作状态及操作方法。

2）在教师指导下，弄清钻床电器元器件安装位置及走线情况；结合机械、电气、液压几方面相关的知识，搞清钻床电气控制的特殊环节。

3）在 Z35 摇臂钻床上人为设置自然故障。

4）教师示范检修。步骤如下：

①用通电试验法引导学生观察故障现象。

②根据故障现象，依据电路图用逻辑分析法确定故障范围。

③采用正确的检查方法，查找故障点并排除故障。

④检修完毕，进行通电试验，并做好维修记录。

⑤由教师设置让学生事先知道的故障点，指导学生如何从故障现象着手进行分析，逐步引导学生采用正确的检修步骤和检修方法。

⑥教师设置故障，由学生检修。

(3) 注意事项。

1）熟悉 Z35 型摇臂钻床电气线路的基本环节及控制要求；弄清电气与执行部件如何配合实现某种运动方式；认真观摩教师的示范检修。

2）检修所用工具、仪表应符合使用要求。

3）不能随意改变升降电动机原来的电源相序。

4）排除故障时，必须修复故障点，但不得采用元器件代换法。

5）检修时，严禁扩大故障范围或产生新的故障。

6）带电检修，必须有指导教师监护，以确保安全。

4.2.5 材料清单

材料清单见表 4-6。

表 4-6 材料清单

序号	名称	数量
1	XK-SX2C 型高级维修电工实训台	1 台
2	维修电工实训组件（XKDT11）	1 个

续表 4-6

序号	名　　称	数量
3	维修电工实训组件（XKDT12A）	1个
4	三相异步电动机	1台
5	跨接线	若干

4.2.6 任务实施

Z35 型摇臂钻床电气线路安装、调试任务书见表 4-7。

表 4-7　Z35 型摇臂钻床电气线路安装、调试任务书

Z35 型摇臂钻床电气线路安装、调试任务书					
1. 根据要求完成所需器件清点并完成下表					
序号	器件名称	器件数量	器件参数	器件测量	备注
2. 工作原理描述					
3. 电路接线图的设计绘制					

续表 4-7

Z35 型摇臂钻床电气线路安装、调试任务书
4. 操作过程记录
5. 通电试车过程记录
6. 任务小结

4.2.7 任务评价

Z35 型摇臂钻床电气线路安装、调试评分表见表 4-8。

表 4-8　Z35 型摇臂钻床电气线路安装、调试评分表

Z35 型摇臂钻床电气线路安装、调试评分表					
器件清点和测量（10 分）					
序号	重点检查内容	评分标准	分值	得分	备注
1	器件清点	器件清点错误一项扣 1 分	5		
2	器件测量	器件测量错误一项扣 1 分	5		
小　　计					
电路图设计（10 分）					
序号	重点检查内容	评分标准	分值	得分	备注
1	主电路设计	主电路错误一处扣 1 分	5		
2	控制电路设计	控制电路设计错误一处扣 1 分	5		
小　　计					
工作原理描述（10 分）					
序号	重点检查内容	评分标准	分值	得分	备注
1	主电路工作原理	是否完整酌情	5		
2	控制电路工作原理	是否完整酌情	5		
小　　计					

续表 4-8

Z35 型摇臂钻床电气线路安装、调试评分表

线路施工（30 分）

序号	重点检查内容	评分标准	分值	得分	备注
1	导线处理是否正确	错误一处扣 1 分	10		
2	导线安装是否牢固	错误一处扣 1 分	10		
3	线路是否正确	错误一处扣 1 分	10		
		小　计			

通电试车（30 分）

序号	重点检查内容	评分标准	分值	得分	备注
1	零压保护功能	功能是否实现	5		
2	主轴控制功能	功能是否实现	5		
3	摇臂升降控制功能	功能是否实现	10		
4	立柱松紧控制功能	功能是否实现	10		
		小　计			

职业素养（10 分）

序号	重点检查内容	评分标准	扣分	得分	备注
1	带电操作	一次扣 5 分扣完为止			
2	规范操作	一次扣 1 分扣完为止			
3	团队合作	酌情			
4	工位整洁	酌情			
		总　计			

4.2.8 故障的排除

故障排除记录单见表 4-9。

表 4-9 故障排除记录单

故障排除记录单		
故障 1	故障现象	
	故障原因	
	故障的排除过程	

续表 4-9

故障排除记录单			
故障 2	故障现象		
	故障原因		
	故障的排除过程		
故障 3	故障现象		
	故障原因		
	故障的排除过程		
故障 4	故障现象		
	故障原因		
	故障的排除过程		

4.2.9 任务拓展

电气设备在运行过程中出现的故障，有些可能是由于操作使用不当、安装不合理或维修不正确等人为因素造成的，称为人为故障。而有些故障则可能是由于电气设备在运行时过载、机械振动、电弧的烧损、长期动作的自然磨损、周围环境温度和湿度的影响、金属屑和油污等有害介质的侵蚀以及电器元器件的自身质量问题或使用寿命等原因而产生的，称为自然故障。显然，如果加强对电气设备的日常检查、维护和保养，及时发现一些非正常因素，并给予及时的修复或更换处理，就可以将故障消灭在萌芽状态，防患于未然，使电气设备少出甚至不出故障，以保证工业机械的正常运行。

电气设备的日常维护保养包括电动机和控制设备的日常维护保养。

(1) 电动机的日常维护保养。

1) 电动机应保持表面清洁，进、出风口必须保持畅通无阻，不允许水滴、油污或金属屑等任何异物掉入电动机的内部。

2）经常检查运行中的电动机负载电流是否正常，用钳形电流表查看三相电流是否平衡，三相电流中的任何一相与其三相平均值相差不允许超过10%。

3）对工作在正常环境条件下的电动机，应定期用绝缘电阻表检查其绝缘电阻；对工作在潮湿、多尘及含有腐蚀性气体等环境条件的电动机，更应该经常检查其绝缘电阻。三相380V的电动机及各种低压电动机；其绝缘电阻至少为 0.5MΩ 方可使用。高压电动机定子绕组绝缘电阻为1MΩ/kV；转子绝缘电阻至少为 0.5MΩ，方可使用。若发现电动机的绝缘电阻达不到规定要求时，应采取相应措施处理后，使其符合规定要求，方可继续使用。

4）经常检查电动机的接地装置，使之保持牢固可靠。

5）经常检查电源电压是否与铭牌相符，三相电源电压是否对称。

6）经常检查电动机的温升是否正常。

7）经常检查电动机的振动、噪声是否正常，有无异常气味、冒烟、启动困难等现象。一旦发现，应立即停车检修。

8）经常检查电动机轴承是否有过热、润滑脂不足或磨损等现象，轴承的振动和轴向位移不得超过规定值。轴承应定期清洗检查，定期补充或更换轴承润滑脂（一般一年左右）。

9）对绕线型转子异步电动机，应检查电刷与滑环之间的接触压力、磨损及火花情况。当发现有不正常的火花时，需进一步检查电刷或清理滑环表面，并校正电刷弹簧压力。刷握和滑环间应有 2~4mm 间距；电刷与刷握内壁应保持 0.1~0.2mm 游隙；对磨损严重者需更换。

10）对直流电动机应检查换向器表面是否光滑圆整，有无机械损伤或火花灼伤。若沾有碳粉、油污等杂物，要用干净柔软的棉布蘸酒精擦去。换向器在负荷下长期运行后，其表面会产生一层均匀的深褐色的氧化膜，这层薄膜具有保护换向器的功效，切忌用砂布磨去。但当换向器表面出现明显的灼痕或因火花烧损出现凹凸不平的现象时，则需要对其表面用零号砂布进行细心的研磨或用车床重新车光，而后再将换向器片间的云母下刻 1~1.5mm 深，并将表面的毛刺、杂物清理干净后，方能重新装配使用。

11）检查机械传动装置是否正常，联轴器、带轮或传动齿轮是否跳动。

12）检查电动机的引出线是否绝缘良好、连接可靠。

（2）控制设备的日常维护保养。

1）电气柜的门、盖、锁及门框周边的耐油密封垫均应良好。门、盖应关闭严密，柜内应保持清洁，不得有水滴、油污和金属屑等进入电气柜内，以免损坏电器造成事故。

2）操纵台上的所有操纵按钮、主令开关的手柄，信号灯及仪表护罩都应保持清洁完好。

3）检查接触器、继电器等电器的触头系统吸合是否良好，有无噪声、卡住或迟滞现象，触头接触面有无烧蚀、毛刺或穴坑；电磁线圈是否过热；各种弹簧弹力是否适当；灭弧装置是否完好无损等。

4）试验位置开关能否起位置保护作用。

5）检查各电器的操作机构是否灵活可靠，有关整定值是否符合要求。

6）检查各线路接头与端子板的连接是否牢靠，各部件之间的连接导线、电缆或保护导线的软管，不得被冷却液、油污等腐蚀，管接头处不得产生脱落或散头等现象。

7）检查电气柜及导线通道的散热情况是否良好。

8）检查各类指示信号装置和照明装置是否完好。

9）检查电气设备和工业机械上所有裸露导体件是否接到保护接地专用端子上。

（3）电气设备的维护保养周期。

对设置在电气柜内的电器元器件，一般不经常进行开门监护，主要是靠定期的维护保养，来实现电气设备较长时间的安全稳定运行。其维护保养的周期，应根据电气设备的结构、使用情况以及环境条件等来确定。一般可采用配合工业机械的一、二级保养同时进行其电气设备的维护保养工作。

4.2.10 任务小结

以 Z35 型摇臂钻床电气线路安装、调试与故障检修训练为例，进一步了解机床电气故障检修及修复注意事项，检修中应分析查明产生故障的根本原因；避免扩大故障；找出故障点后，一定要针对不同故障情况和部位，采取正确的修复方法，不要轻易采用更换电器元器件和补线等方法，更不允许轻易改动线路或更换规格不同的电器元器件，以防止产生人为故障。电气故障修复完毕，需要通电试运行时，应与操作者配合，每次排除故障后，应及时总结经验，并做好维修记录。

参 考 文 献

[1] 曹士勇. 低压电器故障检查与维修方法研究 [J]. 科技风, 2019.
[2] 王涟漪. 对低压电器设备实施状态监测维修模式的探讨 [J]. 内燃机与配件, 2018.
[3] 吕玉明. 维修电工技能训练 [M]. 北京：北京师范大学出版社, 2016.
[4] 王兵. 维修电工国家职业鉴定指南 [M]. 北京：电子工业出版社, 2012.
[5] 杨宗强. 电工维修技能一本通 [M]. 北京：化学工业出版社, 2019.
[6] 韩雪涛. 电工维修全覆盖 [M]. 北京：电子工业出版社, 2019.
[7] 曹金洪. 新编实用电工手册 [M]. 天津：天津科学技术出版社, 2018.
[8] 任清晨. 电气控制柜设计制作 [M]. 北京：电子工业出版社, 2014.
[9] 白公. 维修电工技能手册 [M]. 北京：机械工业出版社, 2017.
[10] 朱照红. 维修电工基本技能 [M]. 北京：中国劳动社会保障出版社, 2019.
[11] 李树海. 电工低压运行维修 [M]. 北京：化学工业出版社, 2011.
[12] 秦钟全. 低压电工上岗技能一本通 [M]. 北京：化学工业出版社, 2012.
[13] 孙洋. 精选实用电工电路 300 例 [M]. 北京：化学工业出版社, 2019.
[14] 秦钟全. 低压电工实用技能全书 [M]. 北京：化学工业出版社, 2017.
[15] 王建华. 电气工程师手册 [M]. 北京：机械工业出版社, 2019.
[16] 张白帆. 低压成套开关设备的原理及其控制技术 [M]. 北京：机械工业出版社, 2019.
[17] 张晓江. 电机及拖动基础 [M]. 北京：机械工业出版社, 2016.
[18] 唐介. 电机与拖动 [M]. 北京：高等教育出版社, 2020.
[19] 陈宝玲. 电机与电控实训教程 [M]. 北京：北京师范大学出版社, 2014.
[20] 程周. 电机拖动与电控技术 [M]. 北京：电子工业出版社, 2013.